Lecture Notes in Computer Science   12655

More information about this subseries at http://www.springer.com/series/7408

Sridutt Bhalachandra · Sandra Wienke ·
Sunita Chandrasekaran ·
Guido Juckeland (Eds.)

Accelerator Programming Using Directives

7th International Workshop, WACCPD 2020
Virtual Event, November 20, 2020
Proceedings

 Springer

Editors
Sridutt Bhalachandra (iD)
Lawrence Berkeley National Laboratory
Berkeley, CA, USA

Sunita Chandrasekaran (iD)
University of Delaware
Newark, DE, USA

Sandra Wienke (iD)
RWTH Aachen University
Aachen, Germany

Guido Juckeland (iD)
Helmholtz-Zentrum Dresden-Rossendorf
Dresden, Germany

ISSN 0302-9743   ISSN 1611-3349   (electronic)
Lecture Notes in Computer Science
ISBN 978-3-030-74223-2   ISBN 978-3-030-74224-9   (eBook)
https://doi.org/10.1007/978-3-030-74224-9

LNCS Sublibrary: SL2 – Programming and Software Engineering

This Springer imprint is published by the registered company Springer Nature Switzerland AG
The registered company address is: Gewerbestrasse 11, 6330 Cham, Switzerland

Preface

The course of high-performance computing (HPC) system architecture is at crossroads. Heterogeneous systems like the current Summit and Sierra, and the up-and-coming systems such as Perlmutter, Frontier, and El Capitan, all have graphics processing units (GPUs). On the other hand, ARM-based systems such as GW4 Isambard and the future Isambard 2 and Fugaku are relatively homogeneous. Nonetheless, system performance relies heavily on acceleration through existing ever-improving vector units in the homogeneous systems or dedicated acceleration units in heterogeneous systems.

With increasing complexity to exploit the maximum available parallelism, the importance of sophisticated programming approaches that can handle performance, scalability, and portability is increasing. Programmers, especially, prefer to keep a single code base to help ease maintenance and avoid the need to debug multiple versions of the same code. In the literature, it has been shown that the abstraction can be raised at different levels - at the high level using directives and frameworks or at a relatively lower level by language modifications.

Software abstraction-based programming models such as OpenMP and OpenACC have been serving this purpose over the past several years and are likely to represent one path forward. These programming models address the 'X' component in a hybrid MPI+X programming approach by providing programmers high-level directives and delegating some burden to the compiler. Such programming paradigms have played a decisive role in establishing heterogeneous node architectures as a valid choice for a multitude of HPC workloads. In addition, frameworks like Kokkos and Raja, along with modifications to the language, are trying to help improve the performance as well as portability.

These proceedings contain the papers accepted for presentation at the 7th Workshop on Accelerator Programming using Directives (WACCPD 2020) held on November 13, 2020. WACCPD is one of the major forums for bringing together users, developers, and the software and tools community to share knowledge and experiences when programming emerging complex parallel computing systems (https://www.waccpd.org).

Like in the previous years, the workshop highlighted improvements to the state of the art through the accepted papers and prompted discussion through keynotes that drew the community's attention to key areas that will facilitate the transition to accelerator-based high-performance computing (HPC). The workshop aimed to showcase all aspects of heterogeneous systems, discussing innovative high-level language features, lessons learned while using directives to migrate scientific legacy code to parallel processors, compilation, and runtime scheduling techniques, among other subjects.

The WACCPD 2020 workshop received seven submissions out of which five were accepted to be presented at the workshop and published in these proceedings. The Program Committee of the workshop comprised 27 members spanning universities, national laboratories, and several industries. Each paper received a minimum of six single-blind reviews. Similar to WACCPD 2019, we encouraged all authors to add the Artifact Description (AD) to their submissions and make their code and data publicly available (e.g. on GitHub, Zenodo, Code Ocean, etc.) in support of the reproducibility initiative. Of the five accepted papers, 40% had reproducibility information and these manuscripts are highlighted with an 'artifacts available' logo in this book.

The program co-chairs invited Prof. Mary Hall from the University of Utah to give a keynote address on "Achieving Performance Portability for Extreme Heterogeneity." Mary Hall is the Director of the School of Computing at the University of Utah. Her research focus brings together compiler optimizations and performance tuning targeting current and future high-performance architectures on real-world applications. Professor Hall is an IEEE Fellow, an ACM Distinguished Scientist, and a member of the Computing Research Association Board of Directors. She actively participates in mentoring and outreach programs to encourage the participation of groups underrepresented in computer science.

Nicholas Malaya from AMD gave an invited talk titled "Enabling Portable Directive-Based Programming at Exascale." Nicholas Malaya is a computational scientist at AMD Research, and is AMD's technical lead for the Frontier and El Capitan Centers of Excellence (COEs). These COEs are focused on close collaborations between AMD, DOE, and HPE to ensure application readiness, so that key workloads can run on the computers from Day-1 of machine deployment. Nick's research interests include Exascale Computing, CFD, Bayesian Inference, and Machine Learning.

Usually, the text of the preface focuses on factual content only. However, 2020 was (unfortunately) different – we feel that we cannot leave this unmentioned: the COVID-19 pandemic hit the world really hard in 2020. It has affected the daily lives of all of us and cost way too many lives. To limit the spread of the virus, the most important action has become social distancing. With that, big events were cancelled all over the world. However, in the HPC community, we have been rather lucky to be able to "easily" switch to solely digital and virtual conference formats. To this end, Supercomputing 2020 was held online, and so was WACCPD 2020 for the first time in its seven-year history. Thanks to all of you that contributed to this success! Hopefully, we will be able to meet in-person again next time. Stay tuned!

February 2021

Sridutt Bhalachandra
Sandra Wienke
Sunita Chandrasekaran
Guido Juckeland

Organization

Steering Committee

Barbara Chapman   Stony Brook University, USA
Duncan Poole   OpenACC, USA
Jeffrey Vetter   Oak Ridge National Laboratory, USA
Kuan-Ching Li   Providence University, Taiwan
Oscar Hernandez   Oak Ridge National Laboratory, USA

General Chairs

Sunita Chandrasekaran   University of Delaware, USA
Guido Juckeland   Helmholtz-Zentrum Dresden-Rossendorf, Germany

Program Chairs

Sridutt Bhalachandra   Lawrence Berkeley National Laboratory, USA
Sandra Wienke   RWTH Aachen University, Germany

Publicity Chair

Neelima Bayyapu   National Institute of Technology Karnataka, India

Web Chair

Shumpei Shiina   University of Tokyo, Japan

Program Committee

Daniel Abdi   National Oceanic and Atmospheric Administration, USA
James Beyer   Nvidia Corporation, USA
Maciej Cytowski   Pawsey Supercomputing Center, Australia
Christopher Daley   Lawrence Berkeley National Laboratory, USA
Joel Denny   Oak Ridge National Laboratory, USA
Johannes Doerfert   Argonne National Laboratory, USA
Millad Ghane   Samsung Semiconductor Inc., USA
Priyanka Ghosh   Washington State University, USA
Haoqiang Jin   NASA Ames Research Center, USA
Ronan Keryell   Xilinx, USA
Jeongnim Kim   Intel, USA
John Leidel   Tactical Computing Laboratories LLC, USA

Ron Lieberman	AMD, USA
Kelvin Li	IBM Corporation, Canada
Meifeng Lin	Brookhaven National Laboratory, USA
Chun-Yu Lin	National Center for High-Performance Computing, Taiwan
Zhao Liu	Tsinghua University, China; National Supercomputing Center, Wuxi, China
Stephen Lecler Olivier	Sandia National Laboratory, USA
Arpith Jacob	Google, USA
Thomas Schwinge	Mentor Graphics, Germany
Bharatkumar Sharma	Nvidia Corporation, Bengaluru, India
Ray Sheppard	Indiana University, USA
Gregory Stoner	Intel, USA
Cheng Wang	Microsoft Corporation, USA
Michael Wolfe	Nvidia Corporation, USA
Rengan Xu	Dell EMC, USA
Charlene Yang	Lawrence Berkeley National Laboratory, USA

Contents

OpenMP

Evaluating Performance Portability of OpenMP for SNAP on NVIDIA, Intel, and AMD GPUs Using the Roofline Methodology

Neil A. Mehta[1]([✉]), Rahulkumar Gayatri[1], Yasaman Ghadar[2], Christopher Knight[2], and Jack Deslippe[1]

[1] NERSC, Lawrence Berkeley National Laboratory, Berkeley, USA
neilmehta@lbl.gov
[2] Argonne National Laboratory, Lemont, USA

Abstract. In this paper, we show that OpenMP 4.5 based implementation of TestSNAP, a proxy-app for the Spectral Neighbor Analysis Potential (SNAP) in LAMMPS, can be ported across the NVIDIA, Intel, and AMD GPUs. Roofline analysis is employed to assess the performance of TestSNAP on each of the architectures. The main contributions of this paper are two-fold: 1) Provide OpenMP as a viable option for application portability across multiple GPU architectures, and 2) provide a methodology based on the roofline analysis to determine the performance portability of OpenMP implementations on the target architectures. The GPUs used for this work are Intel Gen9, AMD Radeon Instinct MI60, and NVIDIA Volta V100.

Keywords: Roofline analysis · Performance portability · SNAP

1 Introduction

Six out of the top ten supercomputers in the list of Top500 supercomputers rely on GPUs for their compute performance. The next generation of supercomputers, namely, Perlmutter, Aurora, and Frontier, rely primarily upon NVIDIA, Intel, and AMD GPUs, respectively, to achieve their intended peak compute bandwidths, the latter two of which will be the first exascale machines. The CPU, also referred to as the host and the GPU or device architectures that will be available on these machines are shown in Table 1.

The diversity in the GPU architectures by multiple vendors has increased the importance of application portability. A wide range of programming frameworks such as, Kokkos, [1] SYCL, [2] and HIP [3] have risen to address this challenge. These languages provide a single front-end for application developers to express parallelism in their codes while the frameworks provide an optimized backend

© Springer Nature Switzerland AG 2021
S. Bhalachandra et al. (Eds.): WACCPD 2020, LNCS 12655, pp. 3–24, 2021.
https://doi.org/10.1007/978-3-030-74224-9_1

Table 1. CPUs and GPUs on upcoming supercomputers.

System	Perlmutter	Aurora	Frontier
Host	AMD Milan	Intel Xeon Sapphire Rapids	AMD EPYC Custom
Device	NVIDIA A100	Intel Xe Ponte Vecchio	AMD Radeon Instinct Custom

implementation on the chosen architecture. However, these programming models require an extensive rewrite of the application codes in C++, including the GPU kernels. Compiler directive-based programming models, such as OpenMP and OpenACC, present an attractive alternative for their ease of use and non-intrusive approach to parallelizing applications. OpenMP has been a popular compiler directive-based programming framework for CPUs and, OpenMP 4.0 onward has included directives that allow application developers to offload blocks of code onto GPUs for execution. OpenMP 4.5 and OpenMP 5.0 have increased the number of directives that will enable effective utilization of the available GPU resources. Compilers such as LLVM/Clang, XL (IBM), Cray, and GCC have already provided backend implementations to offload OpenMP directives on NVIDIA GPUs. Intel and AMD compilers are committed to supporting OpenMP 5.0 directives on their respective GPUs. Meanwhile, NVIDIA has a contract with NERSC to support a subset of OpenMP 5.0 directives for its compiler on the upcoming Perlmutter supercomputer, demonstrating long term investment in supporting OpenMP.

In this paper, we present an OpenMP 4.5 based implementation for the Spectral Neighborhood Analysis Potential (SNAP) module in LAMMPS. [4] Test-SNAP is a stand-alone proxy app for SNAP that can be run independently of LAMMPS and is written in C++. While we have developed Kokkos, CUDA, and HIP versions of TestSNAP that we could have used for this profiling study, the wider use of OpenMP and its support by the GPU vendors makes it the perfect candidate for this study. The goal of this work was to create and test a single source-code implementation that can be compiled and scheduled on the NVIDIA, Intel, and AMD GPUs.

Application "Portability" implies the ability to compile and execute a single source code on multiple architectures. "Performance Portability" includes the ability to efficiently utilize available resources on the said architectures. A more formal definition states that a code can be considered "performance portable" is it consistently achieves consistent ratio of time-to-solution with the best time-to-solution on each platform with minimal platform specific changes to the code. In our study, the use of OpenMP 4.5 ensures that no platform specific changes are required. However, because GPUs from various vendors have different compute architectures, the "time-to-solution" is an inefficient metric for comparison. To assess the efficiency of an application on the target hardware, we have used the roofline analysis to test our OpenMP implementation of TestSNAP by comparing its arithmetic intensity (AI) with the peak achievable AI of the hardware.

We have compiled and executed TestSNAP on testbeds for each of the supercomputers mentioned above, i.e., Perlmutter, Aurora, and Frontier. Testbeds contain intermediary hardware that will fall somewhere between the Summit

Table 2. CPUs and GPUs available on test beds.

Test bed	Cori-GPU	JLSE	Tulip
Host	Intel Skylake	Intel Xeon	AMD EPYC
Device	NVIDIA V100	Intel Gen9	AMD MI60

and exascale systems in terms of power and capabilities. The GPU racks on Cori at NERSC, the Iris node on Joint Laboratory for System Evaluation (JLSE) at Argonne National Lab, and the Hewlett Packard Enterprise built Cray Tulip machine serve as testbeds for Perlmutter, Aurora, and Frontier machines, respectively. GPUs available on each testbed are shown in Table 2. Even though, Intel's Gen9 GPU will not be used on the upcoming HPC machines, it does serve as a good platform to test performance portability of the code on the upcoming next generation discrete GPUs from Intel.

2 OpenMP Offload Implementation of TestSNAP

SNAP is an interatomic potential provided as a component of the LAMMPS MD toolkit. [4] When using the SNAP model, the potential energy of each atom is evaluated as a sum of weighted bispectrum components. The bispectrum, also known as the descriptor, describes the positions of neighboring atoms and the local energy of each atom based on its location for a given structural configuration. This bispectrum is represented by its components, which are used to reproduce the local energy [5]. The neighboring atom positions are first mapped over a three-dimensional sphere using the central atom as the origin to generate the bispectrum components. The mapping ensures that the bispectrum components are dependent on the position of the central atom and three neighboring atoms. Next, we calculate the sum over a product of elements of Wigner D-matrix, a smoothing function, and the element dependent weights. Because this product is not invariant under rotation, we modify it by multiplying with coupling coefficients, analogous to Clebsch-Gordan coefficients for rotations on the 2-sphere, to generate the bispectrum components. The band limit for bispectrum components is set by $\mathbf{J}$, which determines how many and which bispectrum components are used for the simulation. We do not provide a detailed discussion on the SNAP algorithm since it is not in the scope of this paper. Instead, we provide the reader with the implementation details of TestSNAP, the proxy-app for SNAP, since they are necessary to understand it's OpenMP 4.5 implementation. The SNAP algorithm is explained in [6] by the original authors Thompson, *et al.*

2.1 Refactoring Routines for GPUs

The pseudo-code for TestSNAP is shown in Listing 1.1. Each of the compute routines shown in Listing 1.1 iterate over the bispectrum components and store their individual contributions in a 1D array.

Listing 1.1. TestSNAP code

```
1   for(int natom = 0; natom < num_atoms; ++natom)
2   {
3       // build neighbor-list for all atoms
4       build_neighborlist();
5
6       // compute atom specific coefficients
7       compute_U(); //Ulist[idx_max] and Ulisttot[idx_max]
8       compute_Y(); //Ylist[idx_max]
9
10      // for each (atom,neighbor) pair
11      for(int nbor = 0; nbor < num_nbor; ++nbor)
12      {
13          compute_dU(); //dUlist[idx_max][3]
14          compute_dE(); //dElist[3]
15          update_forces()
16      }
17  }
```

idx_max represents the maximum number of bispectrum components and is determined by the value of **J**. TestSNAP problem sizes 2J14, 2J8, and 2J2 represent an idx_max size of 15, 9, and 3, respectively. For all three problem size, we use 2,000 atoms with 26 neighbors for each atom. The three problem sizes denote the number of descriptors used to describe the energy of the atom with respect to its surrounding. Therefore, even though the number of atoms for all three problem sizes remain the same, the number of descriptors used to describe the energy of these atoms ranges as 15, 9, and 3.

The for-loop in line 1 of Listing 1.1 loops over all atoms in the simulation to compute forces in a given time-step. First, a list of neighboring atoms within a certain R_{cut} distance, is generated for each atom inside the routine build_neighborlist. The compute_U routine calculates expansion coefficients for each (atom, neighbor) pair and stores this information in Ulist. The expansion coefficients for each atom are summed over all its neighbors and stored in Ulisttot. Next, the Clebsch-Gordon products for each atom are calculated in the routine compute_Y and stored in Ylist. As a precursor to force calculations, derivatives of expansion coefficients, stored in Ulist, are computed by compute_dU in all 3 dimensions using spherical co-ordinates and stored in dUlist. Using dUlist and Ylist, the force vector for each (atom,neighbor) pair is computed by compute_dE and stored in dElist. Finally, the force on each atom is computed from dElist in update_forces. A correctness check is built-in, which compares the proxy code output against a reference solution.

Based on the strategy used by newer SNAP implementation [7], the basic TestSNAP algorithm discussed above was refactored to prioritize the completion of each stage/routine for all atoms over the completion of all stages/routines for a single atom. In the algorithm shown above, which is based on the older GPU implementation of SNAP, [8], the work of each atom is mapped onto a GPU thread block, and hierarchical parallelism is used to exploit additional parallelism over the neighbor loop and the bispectrum components. However, converting the four major routines, namely, compute_[U,Y,dU,dE] as GPU kernels allow better utilization of GPU resources. In the refactored code, the atom loop is placed inside compute_[U,Y] and similarly, the atom and neighbor loops are placed inside compute_[dU,dE]. As an example, the refactored compute_U, shown in

Listing 1.2 is further refactored into two nested for loops, one to calculate `Ulist` and the other for `Ulisttot`.

Listing 1.2. compute_U

```
1   void compute_U()
2   {
3       compute_uarray();
4       add_uarraytot();
5   }
6   void compute_uarray()
7   {
8       for(int natom = 0; natom < num_atoms; ++natom)
9           for(int nbor = 0; nbor < num_nbor; ++nbor)
10              for(int j = 0; j < idx_max; ++j)
11                  Ulist(natom,nbor,j) = ...
12  }
13  void add_uarraytot()
14  {
15      for(int natom = 0; natom < num_atoms; ++natom)
16          for(int nbor = 0; nbor < num_nbor; ++nbor)
17              for(int j = 0; j < idxu_max; ++j)
18                  Ulisttot(natom,j) += Ulist(natom,nbor,j);
19  }
```

2.2 Use of Multidimensional (MD) Data Structures

One of the disadvantages of refactoring is that it makes it necessary to store the atom and/or neighbor information as individual data structures across all routines. After refactoring, we need to store atom specific information in `Ulisttot` and `Ylist`, and (atom, neighbor) specific information in `Ulist`, `dUlist` and `dElist` arrays. To store this additional information, we create classes that mimic the behavior of multi-dimensional (MD) arrays, such that all elements are stored in a contiguous block of memory to improve memory locality. To achieve this behavior, we have created C++ classes that include a pointer to the contiguous block of memory and the information about the dimensions to calculate indexes of individual elements based on the access pattern.

Listing 1.3. Array2D

```
1   template <class T>
2   struct Array2D
3   {
4       int n1, n2, size;
5       T *dptr;
6
7       Array2D(int in1, int in2)
8               :n1(in1), n2(in2)
9       {
10          size = n1*n2;
11          dptr = new T[size];
12      }
13
14      inline T& operator() (int in1, int in2)
15      {
16          return dptr[in1*n2 + in2];
17      }
18
19      Array2D(const Array2D& p)
20      {
```

```
21          n1 = p.n1; n2 = p.n2; size = 0;
22          dptr = p.dptr;
23        }
24
25        ~Array2D()
26        {
27          if(size && dptr)
28              delete[] dptr;
29        }
30  };
```

A bare bone structure of a 2D class is shown in Listing 1.3. The first and second dimensions of the 2D array are stored as n1 and n2, while size represents the total number of elements, i.e., n1 × n2. dptr points to a contiguous block of memory for size number of elements. The operator overload of () allows us to implement a FORTRAN style indexing for the exact element that is requested. Hence on line 18 of Listing 1.2, the element accessed by Ulisttot will evaluate to Ulisttot.dptr[natom*idx_max + j]. The copy constructor assigns size to zero, which allows us protection against multiple deletions of the same memory block, as shown in the destructor of the class on lines 22–26 of Listing 1.3. Similar to Array2D, Array3D and Array4D classes are created to represent 3D and 4D arrays respectively. Array[2,3,4]D, i.e., ArrayMD classes, are templated over the data type of their elements for generalization. ArrayMD objects of complex-double type are created using a simple structure of two doubles to represent a complex number as shown in line 1 of Listing 1.4. We are aware that there are standard multi-dimensional array classes available through C++ libraries. However, we wanted the ability to control on data storage and array access patterns specific to their usage on CPUs versus on GPUs. Therefore, these classes were created for the purposes of representing MD arrays in TestSNAP and only contain features that are needed by the application.

Listing 1.4. ArrayMD definitions of TestSNAP data structures.

```
1  struct SNAcomplex {double re,im;};
2
3  Array2D<SNAcomplex> Ulisttot(num_atoms,idx_max);
4  Array2D<SNAcomplex> Ylist(num_atoms,idx_max);
5  Array3D<SNAcomplex> Ulist(num_atoms,num_nbor,idx_max);
6  Array4D<SNAcomplex> dUlist(num_atoms,num_nbor,idx_max,3);
```

Data has to be moved from CPU to GPU memory space before distributing the work across GPU threads. We use the map clause in OpenMP to move data between CPU and GPU. Ulist, Ylist, dUlist are only needed on the GPU to store intermediary results between the compute routines. Hence, we use the alloc mapper-type with the map clause to avoid unnecessary memory allocation on CPU. ArrayMD classes are provided with a member function that creates an object without memory allocation, specifically for this purpose. In contrast, dElist is required for computing forces on the CPU after it is updated on the GPU. Therefore, we create a block of memory on the CPU and use the to and from mapper-type for data movement. Listing 1.5 shows how we achieve these two distinct mappings. On line 1 of Listing 1.5, we map Ulist and dElist data structures on to the device using the to mapper-type, which performs a shallow copy of data structures on the device. On line 2, the alloc mapper type is used to allocate a block of memory on the device for `size` number of elements associated with `Ulist`, whereas, in line 3, a deep copy of the memory block pointed by the `dptr` of object `dElist` is performed. We use line 5 to copy the updated `dElist` array back to the CPU.

Listing 1.5. Use of OpenMP directives to `map` data

```
1  #pragma omp target enter data map(to: Ulist, dElist)
2  #pragma omp target enter data map(alloc: Ulist.dptr[0:Ulist.size])
3  #pragma omp target enter data map(to: dElist.dptr[0:dElist.size])
4
5  #pragma omp target exit data map(from: dElist.dptr[0:dElist.size])
```

2.3   Optimizing Routines for OpenMP Offload

In addition to refactoring TestSNAP routines, it is also necessary to understand data access patterns and OpenMP directives for further optimization of Test-SNAP performance. We implemented three optimization strategies, each building on the previous one to make the final TestSNAP version highly performant on all three GPUs.

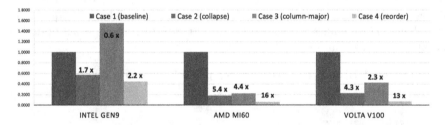

Fig. 1. Code speed-up improvement relative to naive OpenMP 4.5 implementation after array structure and loop access modifications at problem size 2J14.

Code performance is measured using grind-time, which is the average time taken per atom per time-step to complete the force calculation and is calculated in microseconds. The effectiveness of each optimization is measured in terms of,

$$Speed - up = \frac{new\ grind\ time}{naive\ grind\ time}. \tag{1}$$

The *naive grind time* is obtained by running the most trivial GPU parallelization on each architecture. The speed-up measured in this way ensures a fair way of comparing optimization gains specific to each architecture. If the bar is lower than one, it represents speed-up compared to baseline, and a greater than one measurement implies performance degradation. Our results for each of these optimizations are shown in Figs. 1 and 2. The performance plot is divided into three categories, one for each GPU under consideration. Improvement gains due to each optimization step are measured with respect to the naive OpenMP implementation, referred to as Case 1. All optimizations are discussed with the help of add_uarraytot kernel shown in Listing 1.2.

Case 1: We need a baseline to compare the efficiency of our optimizations. Case 1 refers to the naive OpenMP implementation where the atom-loop is distributed across the GPU threads for compute_[U,Y,dU,dE] routines.

Listing 1.6. Atom loop parallelization in add_uarraytot

```
1   void add_uarraytot()
2   {
3   #pragma omp target teams distribute parallel for
4       for(int natom = 0; natom < num_atoms; ++natom)
5           for(int nbor = 0; nbor < num_nbor; ++nbor)
6               for(int j = 0; j < idxu_max; ++j)
7                   ulisttot(natom,j) += ulist(natom,nbor,j);
8   }
```

An example of our naive OpenMP implementation on add_uarraytot is shown in Listing 1.6.

Case 2: Except in compute_Y, each atom loops over its neighbors in all other routines. A logical progression to parallelizing the atom loop is to include the neighbor loop in the parallelization effort wherever possible. We can achieve this by using the collapse clause in OpenMP. An unavoidable consequence of the collapse clause makes it necessary to use atomic operations when updating Ulisttot, as shown in Listing 1.7.

Listing 1.7. Atom and neighbor loop parallelization in add_uarraytot

```
1    void add_uarraytot()
2    {
3    #pragma omp target teams distribute parallel for collapse(2)
4        for(int natom = 0; natom < num_atoms; ++natom)
5            for(int nbor = 0; nbor < num_nbor; ++nbor)
6                for(int j = 0; j < idxu_max; ++j)
7                {
8                    #pragma omp atomic
9                    ulisttot(natom,j) += ulist(natom,nbor,j);
10               }
11   }
```

While atomic calls are expensive, in this case, the benefits of increase in parallelism achieved by looping over the neighbor dimension outweighs the overhead incurred due to atomic operations. Distributing work over the atom and neighbor dimension by the use of collapse clause gave us a 1.7× performance boost on Intel Gen 9, while on AMD and Volta GPUs it gave us a 5.4× and 4.3× performance improvement respectively.

Case 3: One of the most common optimizations on GPUs is the use of column major data access pattern to improve memory coalescing. However, as shown on line 16 of Listing 1.3, we use the row-major style of indexing into the elements of ArrayMD structures which helps to avoid cache thrashing and false sharing on CPUs. Because of the modular design of ArrayMD structure, we can easily modify the operator overload to support column major data access, as shown in Listing 1.8.

Listing 1.8. Column major indexing in Array2D

```
1   inline void operator()(int in1, int in2) {return dptr[in2*n1 + in1];}
```

But this modification does not lead to the intended speed-up. In fact, it leads to performance degradation on all GPUs compared to case 2, as shown in Fig. 1. The reason for this performance degradation is explained in Case 4.

Case 4: The advantage of column major data access on GPUs is the alignment of memory accesses to reduce memory latency on SIMD architectures. In our case this leads to atom dimension being accessed first by consecutive threads. Collapsing the loops makes the index of the innermost loop as the fastest moving index, which implies that the neighbor index becomes the fastest moving index. In order to gain benefit from the column major access pattern, we swap the loop order of atoms and neighbor in each of the routines, as shown on lines 4 and 5 in Listing 1.9.

Listing 1.9. Atom and neighbor loop swap

```
1   void add_uarraytot()
2   {
3   #pragma omp target teams distribute parallel for collapse(2)
4       for(int nbor = 0; nbor < num_nbor; ++nbor)
5           for(int natom = 0; nbor < num_atom; ++natom)
6               for(int j = 0; j < idxu_max; ++j)
7               {
8                   #pragma omp atomic
9                   ulisttot(natom,j) += ulist(natom,nbor,j);
10              }
11  }
```

This allows us to take the advantage of coalesced memory access and gives us the best performance across all 3 GPUs. Speed-ups obtained for problem size 2J8 are similar to those for 2J14, as shown in Fig. 2. Applying case 3 optimization to the 2J8 problem did not degrade the performance to the extent observed in 2J8, which may be because 2J8 problem size relies on smaller ArrayMD structs. The performance gains after case 4 optimizations, although not as high as those for 2J14, are still significant highlighting the efficacy of the applied optimizations.

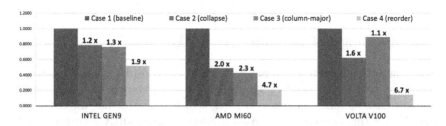

Fig. 2. Code speed-up improvement relative to naive OpenMP 4.5 implementation after array structure and loop access modifications at problem size 2J8.

Compiler maturity plays a significant role in our ability to efficiently map OpenMP directives on GPUs. Because the support for OpenMP directives on

GPUs is still in its early stages, each new version of the compiler can give a significant advantage in terms of new features and increased efficiency of the existing directives. Our OpenMP version of TestSNAP can be successfully compiled and executed on an NVIDIA V100 GPU with the LLVM/10.0 compiler. However, with LLVM/11.0 [9] as well as Intel's®DPC++/C++ (ICX), the code triggers a bug, which results in the compiler being unable to map our SNAcomplex structure, shown in Listing 1.4, on to the device memory. To bypass this bug, we have modified our ArrayMD structures of complex-doubles to structures of doubles with twice the size, such that even and odd indices point to real and imaginary values, respectively. On AMD M160 we have used AOMP version 11.5.1, which is based on LLVM/11.0. The bug reported to LLVM 11.0 has since been fixed in version 12.0 and has been under review by Intel compiler developers. It is important to note that we allowed the compiler to optimize the number of teams and threads when running TestSNAP on all three GPUs. We observed that the compiler optimized teams and threads input always provided better run times compared to runs with manual input.

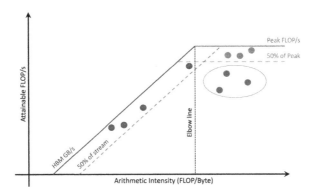

Fig. 3. Schematic of roofline plot, showcasing typical kernel placements.

3   Methodology of Roofline Analysis

Modern computing architectures are varied and are considered guarded proprietary information of the vendor. Therefore, for a fair comparison, the roofline model utilizes a simplified memory model, which assumes that all caches are perfect. Under this assumption, the data flows between DRAM to cache with sufficient bandwidth to not affect performance. Other assumptions include, the communication and computation perfectly overlap, and cores can attain peak floating point operations per second (FLOPs) on local data. These assumptions allow one to measure kernel performance in terms of FLOPs capped by either the peak attainable machine FLOPs or the amount of data that can be moved based on the peak bandwidth throughput.

The measure of how well a kernel can benefit from the data reuse and device bandwidth is quantified by the AI, which is calculated as the number of FLOPs executed per byte of memory transferred to the memory level and is calculated for each level of the memory hierarchy. A roofline plot is formed by plotting attainable FLOPs as a function of AI for a given kernel on a log-log plot. The x- and y-axis represent AI and performance, i.e., FLOPs, respectively. A schematic of a typical roofline plot is shown in Fig. 3. The solid blue line represents the peak attainable bandwidth for a particular memory hierarchy, in this case, the HBM or DRAM of a given machine. No kernels can lie to the left of this line as the attainable FLOP rate will always be bottle-necked by the throughput capacity of the device. At any performance (FLOPs), for a kernel to lie to the left of the bandwidth line, the denominator, i.e., the data transfer rate, will have to be greater than the peak bandwidth of the memory hierarchy. Similarly, the solid green line represents the peak FLOP rate of the machine, which is determined by the machine cycle and is dependent on the compute architecture. The point at which these two bounds meet is known as the "elbow", and the line joining the elbow to the x-axis is called an elbow line. All kernels to the left of this elbow line are termed as "memory-bound" because their performance is strongly affected by the memory bandwidth of the machine. Kernels to the right of the elbow line are classified as "compute-bound" because they are bound by the compute capability of the machine.

A roofline helps determine kernels where optimization efforts are most beneficial. A couple of kernels are shown in blue, green, and red in Fig. 3. Kernels shown in blue lie to the left of the 50% peak bandwidth line. While these kernels have low AI, any additional improvement which increases the FLOPs will lead to a relatively small gain in code performance as these kernels are "memory-bound". Kernels represented by green dots lie above the 50% peak FLOPs rate line and are therefore making good utilization of the machine. In contrast, the kernels shown in red have higher AI than many of the kernels shown in blue, but they have not yet reached 50% of either compute or memory capacity. Optimizing these kernels will provide maximum gains in performance compared to other kernels, which are already capped by either the bandwidth or peak FLOP rate of the machine. The roofline plot also allows one to estimate the kernel performance on future machine architectures. Assuming that an application has a majority of kernels that are "memory-bound", running this application on machines with higher compute capability but the same memory bandwidth will provide only a small improvement in the run-time and vice versa. Ideally, for modern GPUs, where we have more compute power than the memory bandwidth, developers should aspire to make their kernels compute-bound.

4   Results and Discussion

4.1   Profiling Code Performance

To understand the performance difference between LLVM/11.0 and Intel®DPC++/C++ compilers, we have profiled TestSNAP on the Skylake

8180 processor. As shown in Table 3, the step and grind times are similar for LLVM/11.0 and Intel®DPC++/C++ for the serial TestSNAP code on the Skylake processor. The LLVM/11.0 compiler is marginally better, which we suspect may be due to the maturity of the LLVM/11.0 compilers in terms of performance refinement compared to the newly introduced Intel®DPC++/C++. However, for the intent of our comparison, the performance of both compilers on the Skylake processor is considered equal.

Table 3. Comparison of OpenMP offload profiles on GPU measured for the 2J14 problem size for 100 time steps.

Version	Serial (Skylake)		OpenMP offload GPU		
	LLVM/11	ICX	Gen9	MI60	V100
Step time (s/step)	9.7671	9.8669	1.8215	0.1394	0.0565
Grind time (ms/atm-stp)	4.8835	4.9334	0.9107	0.0697	0.0282
compute_U (s)	0.6211	0.6221	0.1975	0.0153	0.0099
compute_Y (s)	7.6839	7.6789	1.2005	0.0748	0.0271
compute_dU (s)	1.2008	1.3363	0.3484	0.0389	0.0155
compute_dE (s)	0.2604	0.2288	0.0741	0.0086	0.0028

Table 4. Comparison of OpenMP offload kernel time loads of top 5 kernels, measured for problem size 2J14, 2000 atoms, and 100 time steps.

Version	Intel Gen9		AMD MI60		NVIDIA V100	
Rank	Time (%)	Kernel	Time (%)	Kernel	Time (%)	Kernel
1	65.65	compute_Y	57.01	compute_Y	45.32	compute_Y
2	19.15	compute_dU	31.53	compute_dU	25.80	compute_dU
3	10.58	compute_U	8.61	compute_U	15.75	compute_U
4	4.02	compute_dE	2.44	compute_dE	8.60	memcpy HtoD
5	0.41	WriteBuffer	0.29	zero_uarraytot	3.96	compute_dE

We use the IProf, ROCProf, and NVProf to profile our OpenMP implementation of TestSNAP on Intel, AMD, and NVIDIA GPUs, respectively. The relative time required by the individual kernels is shown in Table 4. While all three GPUs spend the highest amount of time in compute_Y followed by compute_dU and compute_U, the individual percentages vary. compute_Y is computationally the most expensive kernel followed by compute_dU, and hence they are proportionally the most expensive on each architecture. The initial data movement from device to host on V100, shown in Table 4 as memcpy HtoD contributes 8.6% to the total runtime. Currently, we are unable to obtain the time spent in the data movement on the MI60 GPU using ROCProf, and therefore they are absent in the table. The data movement cost on the Gen9 GPU, represented as `WriteBuffer`, is much lower because of the non-discrete nature of the Gen9 GPU design. `zero_uarraytot` only initializes `Ulisttot` to zero and therefore has very low cost.

It should be noted that the percentage times for the kernels shown in Table 4 are obtained for problem size 2J14 and 100 timesteps. Reducing the number of time steps leads to an increase in the fraction of time spent on data movement. Similarly, reducing the problem size to 2J8 changes the order of kernel time contribution, such that `compute_dU` is most expensive followed by `compute_Y`, `compute_U`, and `compute_dE`. As noted previously, problem size 2J2 is too small to obtain meaningful data for real-world applications. However, we will discuss the roofline results for 2J2 because it highlights some interesting differences between the three GPUs.

4.2   Roofline Analysis of TestSNAP Code

Performance on Intel Gen9 GPU. A high-level building block of Intel Gen 9 GPUs is the slice, and for an Intel Xeon Processor E3-1585 v5 with Iris Pro Graphics P580 (GT4e), which was used in this work, contains 3 GPU slices. Each GPU slice consists of 3 sub-slices, an L3 data cache bank, and shared local memory. A sub-slice has 8 execution units (EUs), each containing 7 threads. Intel processors include fast high bandwidth embedded DRAM (EDRAM) of 128 MB into which the GPU may allocate memory.

We have used Intel® Advisor to collect the relevant metrics necessary to generate the roofline plot and obtain the memory and compute peaks of the Gen9 GPU. Metrics were obtained using the command shown in Listing 1.10.

Listing 1.10. Command to collect the metrics for Intel Gen9

```
1   advixe-cl --collect=roofline --enable-gpu-profiling --project-dir=$PRJ --
        search-dir src:r=$SRC -- ./test_snap.exe -ns 100
```

here, `$PRJ` and `$SRC` denote the locations of user defined project directory into which roofline results are stored and the code source directory, respectively. The roofline plot of TestSNAP running on the Gen9 GPU is shown in Fig. 4 for problem sizes 2J14, 2J8, and 2J2. The numbers in Fig. 4, next to the symbol, correspond to the kernel names shown in the legend. Of note, from the Gen9 GPU roofline plot, while indicating the performance of kernels, it is also possible to obtain details of data flow specific to the Gen9 compute architecture. As a rule, kernel roofline symbols should never cross the memory hierarchy peak bandwidths for which they are measured. For the roofline shown in Fig. 4, the roofline data is generated at the DRAM level for all three problem sizes. However, for 2J14 problem size, the roofline symbols are placed left of the DRAM peak bandwidth line. This indicates that the data movement is not measured across DRAM but across the faster embedded-DRAM or eDRAM, a special feature of the Gen9 GPU.

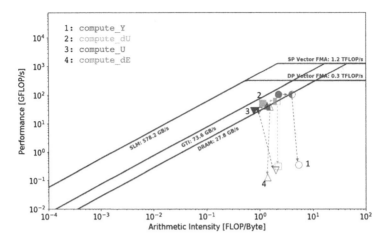

Fig. 4. DRAM roofline plot on **Intel Gen9**. Arrows point from 2J14->2J8->2J2. Problem sizes 2J14, 2J8, and 2J2 are represented by full, half, and open symbols, respectively.GTI and SLM are abbreviations of "Graphics Technology Interface" and "Shared Local Memory", respectively. The SLM is analogous to L3 cache.

All the kernels for the 2J14 and 2J8 problem sizes are bound by the peak bandwidth of DRAM, as indicated by the kernel symbols located close to the DRAM bandwidth line. compute_Y is located close to the elbow created between DRAM bandwidth and DP vector FMA peaks, which represents the region separating "compute" and "memory-bound" regions. This indicates the simultaneous usage of the available compute and memory resources. The other three kernels are not as close to the elbow and are "memory-bound". Finally, based on the location of the 2J14 and 2J8 kernels, it is observed that all kernels are "memory-bound". When running a smaller problem size of 2J8, less data movement is required than 2J14, leading to higher AI for similar performance numbers, leading to a rightward shift of the kernel roofline positions. When running the smallest problem size 2J2, the required number of FLOPs is much less than the data moved, which leads to a large downward shift of the kernel roofline positions, markedly demonstrating poor use of compute resources.

Performance on AMD Radeon Instinct MI60 GPU. We have used the ROC profiling tool to obtain the metrics required for the roofline plot on AMD MI60. ROCProf used in this work is a part of the AMD ROCm version 3.6, which is an open-source code development platform. Unlike Intel Advisor or NVIDIA NSight Compute, we could not obtain the FLOP count of each kernel. Instead, we have used the instruction based roofline model [10] to evaluate the roofline performance of TestSNAP kernels on MI60. To calculate the number of instructions executed, we have used metrics SQ_INSTS_VALU and SQ_INSTS_SALU to obtain the number of vector and scalar instructions issued, respectively. We have used metrics FETCH_SIZE and WRITE_SIZE, to gather read and write data movements, respectively. These metrics are listed in the `input.xml` and provided to ROCProf using the command shown in Listing 1.11. Compute and memory bandwidth peaks were obtained from Richards, *et al.* [11]

Listing 1.11. Command to collect metrics for MI60

```
ROCProf -i input.xml -o roofline.csv ./test_snap.exe -ns 100
```

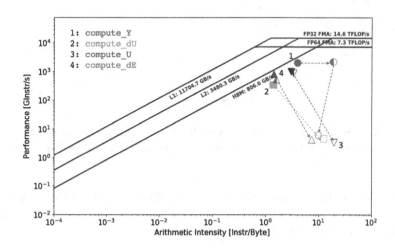

Fig. 5. DRAM roofline plot on **AMD Instinct MI60**. Arrows point from 2J14->2J8->2J2. Problem sizes 2J14, 2J8, and 2J2 are represented by full, half, and open symbols, respectively.

The instructions based roofline data generated using the metric collected at the DRAM level is shown in Fig. 5. `compute_dU`, `compute_U`, and `compute_dE` are all "memory-bound". Similar to the roofline plots discussed previously, kernel `compute_Y` is the most compute-intensive and, therefore, has the highest AI. It is also close to the DRAM peak bandwidth line as well as the compute-bound region, which indicates that `compute_Y` is well-optimized and makes good use of MI60 resources.

Comparing the roofline data of problem size 2J8 with 2J14, except `compute_Y`, all kernels retain their position on the roofline plot. This is because all the necessary data required to run these kernels for problem sizes 2J14 and 2J8 are bound by the DRAM bandwidth, meaning that almost all the required data is fetched from DRAM for both cases. This leads to similar AI and performance numbers for both 2J14 and 2J8. However, this is not the case for `compute_Y`, where the amount of data moved required for instructions executed is lower when running the smaller problem size, 2J8. `compute_Y` relies on the beta coefficients stored in the database files, and at a smaller problem size of 2J8, a lesser number of coefficients are used, and therefore, less data has to be moved. This is definitely the case for all kernels at the smallest problem size 2J2, and consequently, all

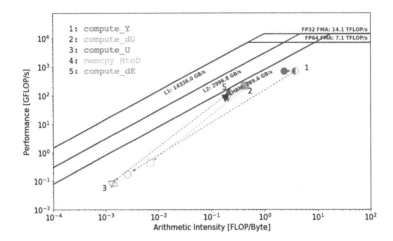

Fig. 6. DRAM roofline plot on **NVIDIA Volta V100**. Arrows point from 2J14->2J8->2J2. Problem sizes 2J14, 2J8, and 2J2 are represented by full, half, and open symbols, respectively.

kernels shift to the "compute-bound" region. The location of the kernels for 2J2 roofline indicates poor use of AMD MI60 GPU resources but shows better utilization at problem size 2J14 and 2J8.

Performance on NVIDIA Volta V100 GPU. The V100 belongs to the Volta family of NVIDIA GPUs and, compared to Intel's Gen9 and AMD's MI60, has been more widely adopted. Metrics necessary to generate the roofline plot were collected using NVIDIA NSight Compute, an interactive kernel profiler. NSight Compute functionality is supported for applications running on NVIDIA GPUs and is provided with CUDA toolkit version 11.0. A total of 11 metrics are collected to obtain the average elapsed time, the number of single and double precision add, multiply, and fused multiply and add (FMA) operations. The data movement across dram, and L2 and L1 caches is tracked for each kernel using metrics `dram__bytes`, `lts__t_bytes`, and `l1tex__t_bytes`, respectively. Metrics necessary for roofline analysis are collected using Listing 1.12.

Listing 1.12. Command to collect metrics for V100

```
1  nv-nsight-cu-cli --metrics $metrics --csv ./test_snap.exe -ns 100 >
      metrics.log
```

where, `$metrics` refers to the metrics discussed above. Compute and memory bandwidth peaks were also obtained from the NSight Compute toolkit.

The DRAM memory roofline of V100, for the three problem sizes, is shown in Fig. 6. Similar to the roofline plots from other GPUs, even for V100, all kernels are positioned in the "memory-bound" regime and are close to the DRAM peak bandwidth line. However, it is possible to observe smaller differences between MI60 and V100 performance. For example, the roofline of `compute_Y`, is comparatively farther away from the DRAM peak bandwidth line in Fig. 6 than in

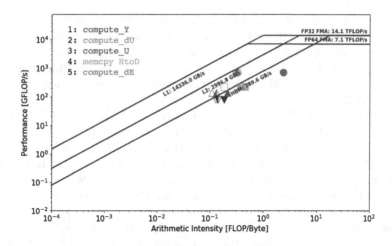

Fig. 7. Hierarchical roofline plot on **Volta V100**, for problem size 2J14. L1, L2, and DRAM performance are represented by open, half, and full symbols, respectively.

Fig. 5. Also, the AI of this kernel is lower than that on MI60. This can potentially be attributed to better communication optimization of TestSNAP kernels on MI60.

This assessment can be made by comparing roofline differences of `compute_Y` for problem sizes 2J14 and 2J8 on these two machines. For problem size 2J8, not as much data has to be moved across the memory levels, which pushes `compute_Y` into "compute-bound" region on MI60, whereas, on V100, the kernel still stays in "memory-bound" region. This suggests that data movement was better for this kernel on MI60 compared to that on V100. Not surprisingly, kernels that are not heavily reliant only on data movement sit closer to the DRAM peak bandwidth line on V100 compared to MI60. The rooflines of kernels for problem size 2J2 are significantly different on the two machines. On MI60, all kernels are located in the "compute-bound" region with relatively higher AI as shown in Fig. 5. In contrast, all the kernels are located in the "memory-bound" region with poor AI, as shown in Fig. 6. Looking at the raw metrics, we can observe that a lot more data is moved across DRAM memory on V100 compared to MI60, and as a consequence, the AI of 2J2 problem size is higher on MI60.

NSight Compute profiler provides an additional level of detail with the ability to capture data transfer not only across DRAM but also across the L2 and L1 cache levels. This data movement can then be used to generate cache specific AI numbers and plot roofline, which can pinpoint data reuse and kernel cache level bounds. Note that the kernel performance is ultimately a minimum of the AI obtained across all memory levels. Roofline models generated in this manner are categorized as hierarchical rooflines.

The L1, L2, and DRAM specific, i.e., the hierarchical roofline plot of Test-SNAP kernels, for problem size 2J14 on V100, is shown in Fig. 7. The noticeable difference in the AI of `compute_Y` indicates that the actual AI of this kernel is not greater than one, but it is approximately 0.3. The proximity of L1 and L2 roofline symbols to the L2 cache peak bandwidth line suggests that `compute_Y` is L2 cache bound. Similarly, except for `compute_U`, the performance of all other profiled kernels is L2 cache bound. For all memory levels, roofline symbols of `compute_U` lie below the DRAM peak bandwidth roof, and therefore, it is considered DRAM bound.

The hierarchical roofline model provides additional details regarding the use of memory hierarchy. In Fig. 7, for `compute_Y`, `compute_dU`, and `compute_dE`, there is a large shift in their AI between DRAM and L2-L1 rooflines. This is a sign of high data reuse and good utilization of memory hierarchy. In contrast, the shift in AI is almost negligible between L2 and L1 cache levels, representing poor utilization of hierarchy, which results from these kernels having to access data from the L2 cache to perform operations. For kernel `compute_U`, because data has to be accessed from DRAM, there is very little shift in AI across L1, L2, and DRAM rooflines, indicating little use of memory hierarchy. The performance of these kernels can be improved by having a larger bandwidth L2 cache and DRAM memory and optimizing and reducing the data movement necessary to execute the kernels.

5   Related Work

Because of the early adoption of OpenMP directives, we were able to learn from the experiences of Vergara Larrea, *et al.* [12] who used OpenMP 4.0 directives to port codes to NVIDIA GPUs. The challenges of using OpenMP 4.5 for performance portability has been documented in detail in work by Gayatri, *et al.* [13] This study laid the groundwork for improving TestSNAP serial version using OpenMP. From this study, it was observed that the collapse clause would be better optimized using the column-major data storage format for 2D and higher dimensional arrays. This early experience helped improve the overall performance of TestSNAP on all tested GPUs. However, for the previous study, the compiler was not as mature in supporting OpenMP offload features. This study demonstrates performance analysis of a real-world application using a mature compiler that is supported by two of the three major GPU architectures.

The roofline model was introduced by Williams, *et al.* [14] in 2009, which made it possible for researchers to measure performance across multiple architectures objectively. Previous works, [15–17] in particular by Yang, *et al.* [18] have been instrumental in developing the theory of the roofline model, tabulating the metrics, and providing a recipe to generate roofline plots. While roofline models have been measured on all three GPU architectures separately, to the best of our knowledge, this is the first time a single application has been analyzed using the roofline model on three GPU architectures with no modifications to the code. This is truly unique because it provides a common standard to measure compiler and GPU architecture improvements.

6 Conclusions and Future Work

In this work, we show that it was possible to create a single source code implementation of TestSNAP using OpenMP 4.5 directives, which is portable across NVIDIA, Intel, and AMD GPUs. To our knowledge, this is the first study the same code was run on three GPU architectures without architecture specific modifications using OpenMP. We also show that standard GPU optimizations such as column-major data access patterns and exploiting more performance by collapsing loops give performance benefits across all GPUs.

TestSNAP run- and grind-times show that the NVIDIA's V100 GPU achieved the highest speed-up with a grind-time of 0.0282 ms/atom-step compared to the serial grind-time of 9.797 ms/atom-step on Intel's Skylake architecture. However, grind-times do not show a complete picture, and this is demonstrated by the roofline models, which show that the TestSNAP kernels are memory-bound on all three GPU architectures. Roofline plots of Gen9, MI60, and V100 indicate that all significant kernels are bound by the DRAM bandwidth. These kernels are positioned in the "memory-bound" region of the roofline plot, and therefore, performance can be improved by changing the algorithm to increase the AI. The ability to collect additional cache level, data movement metrics using CUDA's NSight Compute profiler meant that hierarchical roofline models could be built for TestSNAP on V100 GPU.

As observed from the kernel roofline symbols, the majority of the TestSNAP kernels are "memory-bound". Ideally, kernels should be "compute-bound" and should be closer to the peak compute capacity line. To achieve this, we will work towards better memory access patterns and higher data reuse, particularly for kernel compute_Y, as it has the largest time footprint. Furthermore, we will also work towards better cache utilization to improve the overall AI of the TestSNAP code.

7 Reproducibility

The compiler flags used to compile TestSNAP OpenMP offload on Intel Gen9, AMD MI60, and NVIDIA V100 are provided in Listings 1.13, 1.14, and 1.15, respectively.

Listing 1.13. Compiler flags for Intel Gen9

```
1  icx -O3 -fstrict-aliasing  -Wno-openmp-target -Wall -Wno-unused-variable -
       std=c++11 -qnextgen -fiopenmp -fopenmp-targets=spir64 *.cpp  -o
       test_snap.exe
```

Listing 1.14. Compiler flags for AMD MI60

```
1  clang++ -O3 -fstrict-aliasing  -Wno-openmp-target -Wall -Wno-unused-
       variable -std=c++11 -lm -fopenmp -fopenmp-targets=amdgcn-amd-amdhsa -
       Xopenmp-target=amdgcn-amd-amdhsa -march=gfx906 -ffp-contract=fast *.
       cpp  -o test_snap.exe
```

Listing 1.15. Compiler flags for NVIDIA V100

```
1  clang++ -O3 -fstrict-aliasing  -Wno-openmp-target -Wall -Wno-unused-
       variable -std=c++11 -lm -fopenmp -fopenmp-targets=nvptx64-nvidia-cuda
       --cuda-path=$(CUDA_PATH) -I/$(CUDA_LIB) -ffp-contract=fast *.cpp  -o
       test_snap.exe
```

8 Data Availability Statement

The results shown in this work are reproducible by downloading the code from the git repository: https://github.com/FitSNAP/TestSNAP/tree/OpenMP4.5. However, as mentioned previously, if the code does not compile due to regression test failure, an alternate version of the TestSNAP code without the array of structs is available from github repository: https://github.com/namehta4/TestSNAP/tree/mod_OpenMP4.5. Both these versions have similar compute performance [19].

Summary of the Experiments Reported

We performed roofline analysis of TestSNAP OpenMP 4.5 on Cori GPU, Iris, and Tulip systems at NERSC, JLSE Argonne National Lab, and HPE. We used LLVM/11, Intel ICX, and AOMP compilers for this work. The TestSNAP code version used for this work is available at DOI: 10.6084/m9.figshare.13681816.

Artifact Availability

Software Artifact Availability: All author-created software artifacts are maintained in a public repository under an OSI-approved license.

Hardware Artifact Availability: There are no author-created hardware artifacts.

Data Artifact Availability: There are no author-created data artifacts.

Proprietary Artifacts: None of the associated artifacts, author-created or otherwise, are proprietary.

List of URLs and/or DOIs where artifacts are available:

[breaklines=true, breakanywhere=true]
10.6084/m9.figshare.13681816

Baseline Experimental Setup, and Modifications Made for the Paper

Relevant hardware details: Intel Gen9, AMD MI60, NVIDIA Volta V100

Compilers and versions: clang++ v11.0, Intel ICX, AOMP 11.5-1

Applications and versions: TestSNAP

Libraries and versions: OpenMP 4.5

Key algorithms: Molecular dynamics

Paper Modifications: We refactored the baseline TestSNAP code to optimize for OpenMP offload feature. We have created C++ classes that are comprised of a pointer to a contiguous block of memory and the information about the dimensions to calculate indexes of individual elements based on the access pattern. We also optimized loop structures and data access patterns in the application for offloading to GPUs.

Output from scripts that gathers execution environment information

```
[breaklines=true, breakanywhere=true]
Details regarding baseline experimental setup, and modifications
made for the paper are available at [19].
```

Acknowledgement. The TestSNAP version used in this work is a highly modified variant of the TestSNAP proxy app written by Dr. Aidan Thompson. We would like to thank Drs. Danny Perez, Noah Reddell, and Nicholas Malaya for enabling us access and providing compute resources on the DOE's Cray Tulip machine. We gratefully acknowledge the computing resources provided and operated by the Joint Laboratory for System Evaluation (JLSE) at Argonne National Laboratory. This research used resources of the Argonne Leadership Computing Facility, which is a DOE Office of Science User Facility supported under Contract DE-AC02-06CH11357. We would also like to thank NERSC for providing us with compute resources.

References

1. Edwards, H.C., Trott, C.R., Sunderland, D.: Kokkos: enabling manycore performance portability through polymorphic memory access patterns. J. Parallel Distrib. Comput. **74**(12), 3202–3216 (2014)
2. Reyes, R., Lomüller, V.: SYCL: single-source C++ accelerator programming. In: PARCO, pp. 673–682 (2015)
3. ROCm HIP. ROCm HIP. https://github.com/ROCm-Developer-Tools/HIP
4. Plimpton, S.: Fast parallel algorithms for short-range molecular dynamics. J. Comput. Phys. **117**(1), 1–19 (1995)

5. Bartók, A.P., Payne, M.C., Kondor, R., Csányi, G.: Gaussian approximation potentials: the accuracy of quantum mechanics, without the electrons. Phys. Rev. Lett. **104**(13), 136403 (2010)
6. Thompson, A., Swiler, L., Trott, C., Foiles, S., Tucker, G.: Spectral neighbor analysis method for automated generation of quantum-accurate interatomic potentials. J. Comput. Phys. **285**(1), 316–330 (2015)
7. Plimpton, S., Kohlmeyer, A., Thompson, A., Moore, S., Berger, R.: LAMMPS Stable Release, 3 March 2020. https://zenodo.org/record/3726417#.Xz2NMS2z3Vu
8. Trott, C.R., Hammond, S.D., Thompson, A.P.: SNAP: strong scaling high fidelity molecular dynamics simulations on leadership-class computing platforms. In: Kunkel, J.M., Ludwig, T., Meuer, H.W. (eds.) ISC 2014. LNCS, vol. 8488, pp. 19–34. Springer, Cham (2014). https://doi.org/10.1007/978-3-319-07518-1_2
9. Lattner, C., Adve, V.: LLVM: a compilation framework for lifelong program analysis and transformation. In: CGO, San Jose, CA, USA, pp. 75–88, March 2004
10. Ding, N., Williams, S.: An instruction roofline model for GPUs. In: 2019 IEEE/ACM Performance Modeling, Benchmarking and Simulation of High Performance Computer Systems (PMBS), pp. 7–18. IEEE (2019)
11. Richards, D., et al.: The ECP proxy app team. Quantitative performance assessment of proxy apps and parents. In: LLNL-TR-809403. Report for ECP Proxy App Project Milestone ADCD-504-9. Exascale Computing Project, April 2020
12. Vergara, L.V.G., Wayne, J., Lopez, M.G., Hernández, O.: Early experiences writing performance portable OpenMP 4 codes. In: Proceedings of Cray User Group Meeting, London, England. Cray User Group (2016)
13. Gayatri, R., Yang, C., Kurth, T., Deslippe, J.: A case study for performance portability using OpenMP 4.5. In: Chandrasekaran, S., Juckeland, G., Wienke, S. (eds.) WACCPD 2018. LNCS, vol. 11381, pp. 75–95. Springer, Cham (2019). https://doi.org/10.1007/978-3-030-12274-4_4
14. Williams, S., Waterman, A., Patterson, D.: Roofline: an insightful visual performance model for multicore architectures. Commun. ACM **52**(4), 65–76 (2009)
15. Yang, C., Kurth, T., Williams, S.: Hierarchical roofline analysis for GPUs: accelerating performance optimization for the NERSC-9 Perlmutter system. Concurr. Comput. Pract. Exp. **32**, e5547 (2019)
16. Konstantinidis, E., Cotronis, Y.: A quantitative roofline model for GPU kernel performance estimation using micro-benchmarks and hardware metric profiling. J. Parallel Distrib. Comput. **107**, 37–56 (2017)
17. Ding, N., Williams, S.: An instruction roofline model for GPUs. In: 2019 IEEE/ACM Performance Modeling, Benchmarking and Simulation of High Performance Computer Systems (PMBS), pp. 7–18 (2019)
18. Yang, C., et al.: An empirical roofline methodology for quantitatively assessing performance portability. In: 2018 IEEE/ACM International Workshop on Performance, Portability and Productivity in HPC (P3HPC), pp. 14–23 (2018)
19. Mehta, N., Gayatri, R., Ghadar, Y., Knight, C., Deslippe, J.: Artifact and instructions to generate experimental results for WACCPD conference proceeding 2020 paper: Evaluating Performance Portability of OpenMP for SNAP on NVIDIA, Intel, and AMD GPUs using the Roofline Methodology (2021)

Performance Assessment of OpenMP Compilers Targeting NVIDIA V100 GPUs

Joshua Hoke Davis[1]([⊠]) [iD], Christopher Daley[2], Swaroop Pophale[3],
Thomas Huber[1], Sunita Chandrasekaran[1], and Nicholas J. Wright[2]

[1] University of Delaware, Newark, DE 19716, USA
jhdavis@udel.edu
[2] National Energy Research Scientific Computing Center, Lawrence Berkeley
National Laboratory, Berkeley, CA 94720, USA
[3] Oak Ridge National Laboratory, Oak Ridge, TN 37830, USA

Abstract. Heterogeneous systems are becoming increasingly prevalent. In order to exploit the rich compute resources of such systems, robust programming models are needed for application developers to seamlessly migrate legacy code from today's systems to tomorrow's. Over the past decade and more, directives have been established as one of the promising paths to tackle programmatic challenges on emerging systems. This work focuses on applying and demonstrating OpenMP offloading directives on five proxy applications. We observe that the performance varies widely from one compiler to the other; a crucial aspect of our work is reporting best practices to application developers who use OpenMP offloading compilers. While some issues can be worked around by the developer, there are other issues that must be reported to the compiler vendors. By restructuring OpenMP offloading directives, we gain an 18x speedup for the su3 proxy application on NERSC's Cori system when using the Clang compiler, and a 15.7x speedup by switching max reductions to add reductions in the laplace mini-app when using the Cray-llvm compiler on Cori.

Keywords: Directive-based programming · Performance portability · Heterogeneous systems · OpenMP · GPU · NVIDIA · V100

1 Introduction

Of the 500 supercomputers on the Top500 list, a full thirty percent (150 systems) use many-core technologies, such as NVIDIA Volta GPUs or Intel Xeon Phi many-core co-processors [28]. This is up from 133 systems in the list one year ago. Furthermore, seven of the top ten supercomputers on the latest list use many-core technology. Heterogeneous architectures, those which use co-processors or accelerators in addition to a main processor, are valued for their energy efficiency and promise significant performance gains for applications that can make use of

© Springer Nature Switzerland AG 2021
S. Bhalachandra et al. (Eds.): WACCPD 2020, LNCS 12655, pp. 25–44, 2021.
https://doi.org/10.1007/978-3-030-74224-9_2

them. However, programming for these platforms, and in particular, porting existing applications to these platforms, poses a significant challenge. Scientific programmers look forward to taking advantage of these powerful architectures without having to learn the exact hardware details or make significant changes to their applications, which can often exceed tens of thousands of lines of code.

Numerous programming models and tools exist for programming heterogeneous systems, including CUDA [21], OpenCL [13], and Kokkos [15]. Directive-based models such as OpenACC [23] and OpenMP [24] are popular solutions, as they offer a useful degree of abstraction over various hardware types with a unified interface, and reduce the work needed to accelerate an application, requiring only "hints" or annotations to be added to the compiler. The OpenMP model introduced support for offloading code (with the target directive) to accelerators, co-processors, or many-core processors from version 4.0 (released 2013), and has continued to add and update features through versions 4.5 (released 2015) and 5.0 (released 2018).

To understand the value of offloading support in OpenMP, we highlight the following points from the 2018 NERSC-10 workload analysis: more than eighty percent of the NERSC community uses OpenMP for parallel programming, making OpenMP far and away the most widely-adopted model in use at NERSC, and fifty-one percent of the NERSC workload is already either fully or partially implemented on GPUs [1]. Meanwhile, the 2019 OLCF Operational Assessment indicates that its three allocation programs all used more than 75% of their hours spent on Summit running GPU-enabled jobs, with INCITE reaching 94% GPU-enabled hours [22].

Knowing that heterogeneous architectures are only continuing to grow in popularity, it is critical that users understand the status of the various vendor compilers which support OpenMP offloading. Application developers must be able to make an informed choice of compiler based on which particular offloading features their application uses. Understanding cases in which identical OpenMP directives can show highly variable performance across compilers is essential to making such a decision. And, where compilers exhibit performance differences, understanding the underlying reasons in the implementation for those differences is useful not only for improving the portability of an application between compilers but also for giving specific feedback to vendors about the limitations of their existing implementations.

The main contributions of this work are as follows:

– Identify five benchmarks and proxy applications (mini-apps) which characterize the performance of OpenMP offloading features used by major applications and exhibit performance differences across compilers that are of interest to developers.
– Quantify performance differences across state-of-the-art compilers for the benchmarks and proxy applications selected.
– Explain the observed differences in performance between implementations by using profiling tools and performance metrics, making use of an execution

time decomposition methodology, where needed, to quantify the impacts of kernel launch latency and OpenMP runtime overhead.
- Make recommendations to application developers regarding the best practices for performance portable OpenMP offloading, guided by insights into the causes of slowdowns in kernels derived from real-world applications.

The remainder of this paper is organized as follows. In Sect. 2 we discuss related work in compiler comparison for heterogeneous architectures. In Sect. 3 we describe the five mini-apps selected, as well as the environment and methodology used to test these mini-apps. Section 4 shows the results for each of the five mini-apps, both in terms of general performance and specific insights gained from profiling, and Sect. 5 sets out recommendations to application developers based on the insights gained. Finally, in Sect. 6 we conclude the paper and identify directions for future work.

2 Related Work

Several existing works narrate the use of OpenMP offloading features for many-core processors and accelerators such as GPUs. These include performance analysis of TeaLeaf and CloverLeaf [18], as well as LULESH [2], which uses OpenMP 4.0. Larrea et al. [17] show preliminary lessons learned writing portable code using OpenMP 4.0. Gayatri et al. [10] used a material science application with OpenMP 4.5 to compare and contrast with OpenACC, showing that an unchanged OpenMP GPU version of the code was ill-suited for CPU execution. ExaHyPE, an Exascale Hyperbolic PDE design [30] used a pragma-based GPU parallelization approach for object-oriented code, and documented lessons learned. Several other related works include demonstrating GPU support for OpenMP offloading features in compilers in Flang/Clang [3,25], a proof-of-concept implementation of offloading for FPGA based accelerators [14,26], and an interprocedural statical analysis heuristic at runtime to select optimal grid sizes for offloaded target team constructs [27], among others.

There are publicly available benchmark suites to evaluate heterogeneous application performance, e.g. SPEC-ACCEL [11,12] and Rodinia [6]. The performance of the SPEC-ACCEL benchmark suite was evaluated on multiple platforms using multiple OpenMP offloading and OpenACC compilers by Boehm et al. [4]. Here, the authors reported a list of compilation/runtime errors for individual benchmarks as well as benchmark execution time, however, there was little detail about reasons for the observed performance with different compilers. The Rodinia benchmark suite was used to evaluate OpenMP offloading Unified Memory performance by Mishra et al. [19]. The OpenMP offloading and OpenACC performances of four mini-apps were evaluated across platforms and compilers by Larrea et al. [29]. Larrea et al. described the development coding challenges, portability issues and performance, but did not go into detail about the reasons for poor performance reported. A detailed evaluation of the overhead of different OpenMP compilers was performed by Diaz et al. [20], however, this had a narrow focus on the overhead of individual OpenMP constructs.

In contrast to existing related work, this paper focuses on a set of mini-applications, thus forming a suite of codes using two major systems: NERSC Cori and ORNL Summit. We explore the compatibility of the mini-apps with 7 compilers including 5 OpenMP offloading, 1 OpenACC, and 1 CUDA compiler to quantify and document performance differences across compilers *and* offer recommendations to application developers for usability and best practices for OpenMP offloading compilers.

3   Mini-apps Suite and Experimental Setup

3.1   Mini-apps Suite

The suite is made up of mini-apps chosen for their focus on offloading kernels, diverse characteristics, and their ability to be compiled by all available compilers. The following benchmarks and proxy applications were selected for this paper:

1. **su3** [8] is a matrix-matrix multiply code using complex numbers. It is extracted from MILC (MIMD Lattice Computation), a Lattice QCD (Quantum Chromodynamics) code.
2. **babelStream** [7] is a memory bandwidth benchmark implemented in multiple programming models. It measures the rate of transfer to and from the global device memory with a number of computational kernels, including dot, add, mul, copy, and triad.
3. **laplace** (ported from [5]) is an implementation of an iterative Jacobi method Laplace equation solver, which launches multiple small stencil update kernels and uses the OpenMP reduction clause to check for convergence.
4. **gpp** [9] is a proxy application for the generalized plasmon-pole model from BerkeleyGW, a many-body perturbation theory code. gpp relies on an reduction to compute its final result.
5. **ToyPush** (ported from [16]) is a proxy application for the electron push phase in XGC1, a particle-in-cell simulation code for magnetically-confined fusion plasma. It is similar to laplace in that it launches a large number of short-running kernels.

3.2   Systems and Compilers

All results shown in this paper use NERSC's Cori machine (GPU testbed nodes) and the Summit supercomputer at the Oak Ridge National Laboratory (ORNL). Table 1 shows the hardware details of these systems.

Table 2 shows the compilers tested for each mini-app, where possible. Because PGI support for OpenMP offloading is still under development, PGI was tested using an OpenACC equivalent implementation of each code. Note that the Clang 11 versions used on Cori are both the same in-development version. The Cray Classic compiler (CCE 9.0.0) refers to the Cray C/C++ compiler that uses proprietary Cray compiler technology, in Cray CCE 10.0.0 the C/C++ compilers have been replaced with Cray enhanced LLVM and clang. This not only means

Table 1. Overview of the Cori-GPU and Summit systems.

	Cori-GPU	Summit
Node architecture	Cray CS-Storm 500NX	IBM AC922
Node CPUs	2 × Intel Skylake	2 × IBM Power 9
Available cores per CPU	20 @ 2.40 GHz	21 @ 3.07 GHz
Node GPUs	8 × 16 GB NVIDIA V100	6 × 16 GB NVIDIA V100
CPU-GPU interconnect	PCIe 3.0 switch	NVLink 2.0

that nearly all of the compiler flags are different, but also that the performance will be different. Table 3 shows which of our mini-apps can be compiled and run with which compilers. A status of NI indicates that the mini-app is not implemented in the required programming model for that compiler, while CE and RE indicate compiler and runtime errors, respectively. LLVM's Fortran compiler Flang does not have complete support for OpenMP offloading features, so for the ToyPush application (the sole Fortran app tested) LLVM results cannot be shown.

Throughout this paper, application results are verified whenever the app runs to completion. Each compiler was used with the most aggressive optimization flags enabled, i.e. -Ofast (or equivalent if named differently).

Table 2. Compilers and GPU offloading methods evaluated on the Cori-GPU and Summit systems

Compiler	GPU offload	Cori-GPU version	Summit version
NVCC	CUDA	10.2.89	–
NVIDIA/PGI	OpenACC	20.4	–
Cray CCE	OpenMP	10.0.0 (LLVM version)	–
Cray CCE	OpenMP	9.0.0 (Classic version)	–
IBM XL	OpenMP	–	16.1.1-5
LLVM/Clang	OpenMP	11.0.0-git (#17d8334)	11.0.0-git (#17d8334)
GNU/GCC	OpenMP	–	9.1.0

3.3 Profiling Methods and Tools

Our approach for understanding performance differences across compilers starts from identifying where such performance differences exist. For each mini-app, tested across all compilers it is compatible with (see Table 3), we first record a metric of performance, which varies depending on the nature of the application. For example, su3 has a figure of merit of GFLOPs. For more complex apps such

Table 3. Compatibility of mini-apps with each compiler. (NI: No implementation for required programming model; RE: Runtime Error)

Compiler	su3	babelStr.	laplace	gpp	ToyPush
NVCC (CUDA)	✓	✓	NI	NI	NI
PGI (OpenACC)	✓	✓	✓	✓	✓
Cray-llvm	✓	✓	✓	✓	✓
Cray-classic	✓	✓	✓	✓	✓
XL	✓	✓	✓	✓	✓
Clang	✓	✓	✓	✓	–
GCC	✓	✓	✓	✓	RE

as ToyPush or laplace, execution time was used, while for babelStream, which is memory-bound, we measured memory bandwidth. If the chosen metric for a given application is relatively poor for one compiler compared to others, that indicates this compiler is generating inefficient code.

Knowing which compilers perform poorly for a given application, we use profiling tools to uncover the underlying reasons for such poor performance. The two profiling tools used in this study are nvprof and Nsight Compute, both NVIDIA products. nvprof is a command-line profiler for NVIDIA GPUs, which we use to identify GPU activities and kernels that are most time intensive and collect hardware metrics relating to memory use and instruction counts. Nsight Compute, which has a command-line and graphical component, we use to profile the kernels of an application in-depth. Nsight Compute indicates high-level bottlenecks, creates roofline plots, and features a source analysis view which we use to identify high-latency sections of a kernel in both the original source and generated assembly code.

The choice of metrics to focus on for a particular application depends on our understanding of what the application is computing, as well as the bottlenecks indicated by Nsight Compute. For example, Nsight Compute tells us babelStream's dot kernel is latency bound when compiled with Clang, with low compute and memory utilization. Knowing this, we use source analysis to identify which source lines have the most latency samples, confirming the impact of the OpenMP reduction specifically. Viewing the SASS (Shader Assembly) alongside the source can provide a deeper understanding of where latency specifically arises, such as constant memory load instructions that appear in some codes compiled with Cray-classic.

In other cases, there is less to be learned from the hardware metrics to gain a deep understanding of a kernel, as the application launches many small kernels rather than a few large kernels. In these cases, we expect kernel launch latency and overhead of the OpenMP runtime to be a major cause of performance degradation. The NVTX (NVIDIA Tools Extension) API provides a set of CPU functions to tag parts of software for GPU profiling. With NVTX, we are able to wrap an OpenMP `target` region in an NVTX range, so that nvprof will specifically time the region. The general form of this approach is shown in Listing 1.1, using the code structure of the laplace mini-app (see Sect. 4.3).

```
id1 = nvtxRangeStartA("launch");
#pragma omp target teams distribute parallel for reduction
    (..) collapse(2)
for (i = 1; i <= height; ++i) {
    // stencil update...
}
nvtxRangeEnd(id1);
```

Listing 1.1. NVTX range markers

The nvprof profiler then gives us an average duration of the range as well as the average time spent on kernels and data movement in that region. Assuming no overlap, this breakdown is made up of three parts, as shown in Eq. 1.

$$NVTXRangeTime = GPUTime + CPUTime + DataMovementTime \quad (1)$$

Intuitively, this means that starting with the NVTX range total time reported by nvprof and subtracting the average data movement and GPU kernel time leaves us with the CPU time. This time is accounted for primarily as overhead of the OpenMP runtime.

Note that all profiling and sampling data collected has less than 5% variation between runs. Profiling overhead varies depending on the tool and configuration. Nsight Compute can show 3x–20x slowdowns, while nvprof *without* metric collection shows a minimal slowdown, around 1.1x to 1.2x from our tests. nvprof with metric collection shows a 1.3x to 7x slowdown. All execution times shown are measured *without* profiling tools.

Section 4 will show the results collected for each mini-app across compilers, as well as our insights into causes for performance differences taken from performance metrics and profiling.

4   Results

As described in Sect. 3.3, our investigation starts from identifying which applications show drastic performance differences across compilers. Figure 1 shows, for each application version and compiler, the degree of variance in performance of the tested compilers. The differences between versions shown for each mini-app will be described in the following subsections. Performance for this figure is on relative scale from zero to one, where one represents the performance of the best-performing compiler, using the most appropriate metric for each benchmark. For example, because babelStream is a memory bandwidth benchmark, the memory bandwidth achieved is used as the performance metric for comparing compiler performance, while laplace uses total execution time. The following subsections will describe the differences between versions for each mini-app, and examine the performance variation in detail.

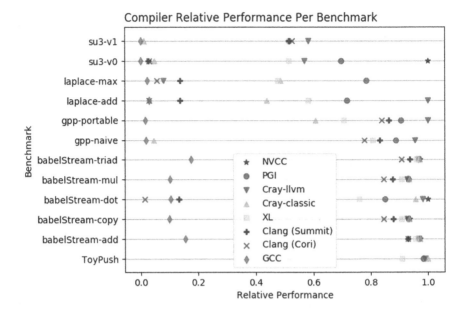

Fig. 1. Relative performance for each mini-app and compiler.

4.1 su3

The su3 mini-app is a matrix-matrix multiply code. Figure 2 shows the GFLOPs per compiler, computed using the observed execution time based on the known number of operations the kernel performs. The theoretical peak performance on NVIDIA V100 GPUs based on the calculated arithmetic intensity of the kernel is 1,270 GFLOPs.

Compared to the CUDA baseline, GCC, Clang and Cray-classic stand out as poorly-performing, with <1%, 3% and 5% of the CUDA performance, respectively. Using Nsight Compute profiling, we attribute Cray-classic performance to small grid size, and therefore poor device utilization as well as latency issues arising from rapid, intense use of global constant memory. Small grid size is compensated for by increasing the number of teams using the num_teams clause. By raising the number of teams from 1200, the Clang-tuned value, to 10000, Cray-classic reaches approximately 240 GFLOPs, or a 4.6x speedup.[1] In comparison, the best-performing compiler (NVCC) used a grid size of 294912, and the worst-performing compiler (GCC) used a grid size of 1600.

To further investigate Clang performance, we examine DRAM transactions for each compiler. Data is collected using the nvprof command-line profiler. According to the DRAM read and write transaction metrics, su3 performs excess

---

[1] This is also a 2.03x speedup compared to the performance of the Cray-classic-chosen default value, which is 81920 teams. Note that Cray-classic ignores num_threads, as it only considers teams and SIMD parallelism.

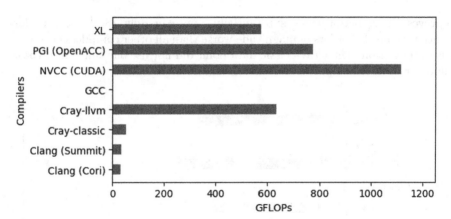

Fig. 2. GFLOPs per compiler for su3-v0. Performance results are obtained on Cori-GPU, except for "XL", "GCC" and "Clang (Summit)" data points.

DRAM data movement when compiled with Clang, over 20x more write trans-actions and 3x more read transactions when compared to the CUDA baseline.

Listing 1.2 shows the OpenMP construct arrangement in su3. According to the Clang documentation, this arrangement of directives, specifically, the inter-leaving for loop between the **teams** and **parallel** constructs, causes Clang to choose the non-SPMD mode for code generation. To test if the use of non-SPMD mode is responsible for elevated DRAM transactions, we modify the OpenMP directive structure of su3, as shown in Listing 1.3. This optimized version removes interleaving code between the **teams** and **parallel** constructs, and manually distributes the loop iterations between teams.

```
1 #pragma omp target teams distribute
2     for(int i = 0; i < total_sites; ++i) {
3 #pragma omp parallel for collapse(3)
4         // 3 for loops...
```

Listing 1.2. OpenMP directives in su3-v0

```
1 #pragma omp target teams
2 #pragma omp parallel
3 {
4     // compute istart, iend for each team ...
5     for(int i = istart; i < iend; ++i) {
6 #pragma omp for collapse(3)
7         // 3 for loops...
```

Listing 1.3. OpenMP directives in su3-v1

With su3-v1's modifications, the DRAM transactions for all compilers except Clang remain approximately the same, while Clang DRAM transactions fall to a level matching the other compilers. Examining the GFLOPs per compiler, as shown in Fig. 3 for su3-v1, shows that this change to the OpenMP directives improves Clang performance substantially, approximately 18x.[1] GCC performance remains 2–3 orders of magnitude worse than all other compilers even with this optimization. The Cray-classic data point did not use our tuned num_teams value, for comparison purposes.

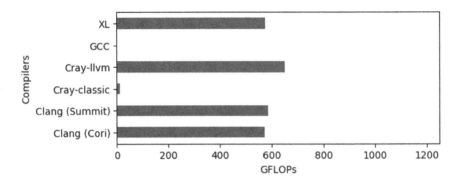

Fig. 3. GFLOPs per compiler for su3-v1. Performance results are obtained on Cori-GPU, except for "XL", "GCC" and "Clang (Summit)" data points.

4.2 babelStream

The babelStream memory bandwidth benchmark uses a number of simple compute kernels to test memory bandwidth, called dot, copy, add, mul, and triad. The dot kernel is unique in that, unlike the other kernels, it uses a reduction clause in its computation. As babelStream is a global device memory bandwidth benchmark, we expect it to be memory-bound, reaching near-peak memory bandwidth (900 GB/s). Figure 4 shows the measured memory bandwidth for the dot, copy, and add kernels for each compiler.

The dot kernel compiled with Clang stands out in Fig. 4 as performing poorly. GCC performance for all babelStream kernels is also relatively poor. Nsight Compute identifies the dot kernel, when compiled with Clang, as latency-bound, rather than bandwidth-bound as expected. Nsight Compute Warp State Analysis points out that the kernel has stall issues, with each warp on average spending 28.7 cycles waiting on a barrier, and Source view shows that there are a large number of barrier latency samples collected on the OpenMP directive that has the reduction clause. Taking into account the lack of similar latency issues on any other babelStream kernel, we infer that these barrier samples must arise due to the introduction of the reduction clause.

---

[1] Note that after these modifications, Clang chooses default num_teams and num_threads values of 128 and 128, which do not perform as well as our tuned values of 1600 and 64 (4.45x speedup with tuned values compared to defaults).

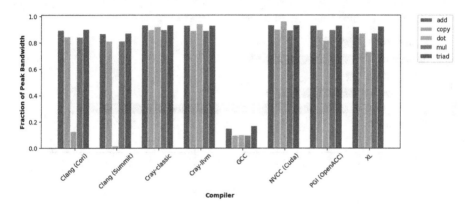

Fig. 4. Fraction of peak memory bandwidth per compiler for babelStream.

4.3 laplace

The laplace mini-app has two features which we suspect to be possible performance impediments: first, it uses a reduction clause to determine if the computation has converged, and second, it executes a large number of short-running kernels, which would increase the impact of the OpenMP runtime overhead and of any kernel launch latency. As described in Sect. 3.3, NVTX range markers allow us to measure the composition of execution time for a given offloading region, following Eq. 1. Figure 5 shows the results of this approach for each compiler.

Cray-llvm performs poorly, due to high GPU time, while Clang and GCC also perform poorly, due to high CPU and GPU time. This high CPU time in Clang indicates a high overhead of the OpenMP runtime, which negatively impacts performance. Clang GPU time is longer on Summit compared to on Cori, which according to Nsight Compute profiling, is caused by elevated (about 10x higher) barrier latency on the max reduction for Clang on Summit, specifically on one move instruction compared to Clang on Cori. This is causing far more warp stalling and thus lower compute and memory utilization.

The most significant limiter of laplace performance is the use of a max `reduction` clause. Add reductions are shown in the babelStream study to be a source of latency issues in Clang, but did not pose a problem for Cray-llvm. To confirm that the max reduction specifically causes the high Cray-llvm GPU Time seen in Fig. 5, we create a version of the laplace app that uses an add reduction rather than a max reduction. Rather than detecting convergence, it merely iterates a fixed number of times, the number of iterations the max reduction version needed to converge. Most compilers show little performance difference between the max and add reduction versions, but Cray-llvm shows a significant difference, a 15.7x speedup in GPU Time using add reduction version.

Profiling with Nsight Compute shows the reasons for this extreme difference in max and add reduction performance in Cray-llvm. Source analysis indicates that the max reduction clause has a large number of Long Scoreboard latency

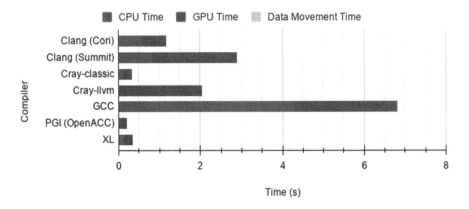

Fig. 5. Breakdown of execution time in laplace. Results are shown in seconds.

samples. Each warp of the kernel spends on average 53.6 cycles stalled waiting for a L1TEX (global memory) operation, meaning that the kernel is using global memory heavily. Further investigation into the assembly generated for the reduction shows the source: an atomic operation on global memory. By comparison, the add reduction in Cray-llvm has far fewer latency samples, and no similar atomic operation.

The difference between the max and add reduction implementations in Cray-llvm can be further confirmed with hardware metrics. nvprof profiling finds that the metrics `atomic_transactions` and `l2_atomic_transactions` are both approximately 2740 times higher for the max reduction version compared to the add reduction version.

Nsight Compute SASS view shows specifically that Cray-llvm uses fewer hardware atomic-add instructions, implying the use of tuned reduction algorithms, e.g. 5-stage hierarchical shuffle-based algorithms. This is not the case for the Clang compiler, which uses a relatively large number of general purpose compare-and-swap atomic instructions. Detailed analysis of the Clang compiler shows a similar count of compare-and-swap atomic instructions for the max and add reductions, implying reuse of the compiler code. The Cray-llvm compiler uses 4 orders of magnitude more atomic instructions than Cray-classic implying use of a general purpose slower code path. Only three compilers, PGI, Cray-classic and XL, generate an efficient max reduction according to Fig. 5.

4.4   gpp

gpp is a larger mini-app, which uses an add reduction to compute its final result. There are two versions tested for gpp: gpp-portable, which includes the default reduction reconfiguration described below, and gpp-naive, which removes that reconfiguration. Measuring execution time of gpp-portable across compilers we observe consistency across compilers, save for GCC. Examining the use of the `reduction` clause in gpp-portable, we see a reconfiguration approach to mitigate the impact of reduction slowdowns in some cases, which explains this generally consistent good performance. To understand the possible benefits of this approach, consider how the reduction would usually be done (i.e., as it is in gpp-naive), shown in a simplified form in Listing 1.4.

```
#pragma omp target teams distribute parallel for simd
    collapse(2) reduction(+:sum)
    for(/* iterate over ngpown */) {
        for(/* iterate over num_bands */) {
            for(/* interate over ncouls */) {
                // compute values...
                sum += computed_value;
            }
        }
    }
```

Listing 1.4. gpp-naive reduction usage

The reduction operator in gpp-naive is placed in the innermost loop, as per usual, so that every iteration of the innermost loop adds to the `reduction` variable. By comparison, gpp-portable moves the reduction operations one loop up, after the innermost loop inside the middle loop. The innermost loop instead sequentially stores the results of the innermost loop in a local variable, which is only reduced after the inner loop is complete. Listing 1.5 demonstrates the approach, simplified.

```
#pragma omp target teams distribute parallel for simd
    collapse(2) reduction(+:sum)
    for(/* iterate over ngpown */) {
        for(/* iterate over num_bands */) {
            double local_sum = 0.0;
            for(/* interate over ncouls */) {
                // compute values...
                local_sum += computed_value
            }
            sum += local_sum;
        }
    }
```

Listing 1.5. gpp-portable reduction usage

Figure 6 compares the execution time of gpp-portable and gpp-naive, and as expected, gpp-naive is generally slower than gpp-portable. gpp-portable, compared to gpp-naive, shows a 1.02x to 1.05x speedup in kernel time for the Clang, Cray-llvm, and PGI compilers, a 13.4x speedup for Cray-classic, and a 0.88x and 0.95x slowdown for XL and GCC (meaning the change harms GCC and XL performance, and only slightly improves performance for other compilers except Cray-classic). GCC is also relatively poorly-performing compared to other compilers. The particularly poor Cray-classic performance on gpp-naive can be attributed to elevated device memory activity, as it shows approximately 3714 times more bytes transferred to and from device memory compared to Clang's version of gpp-naive.

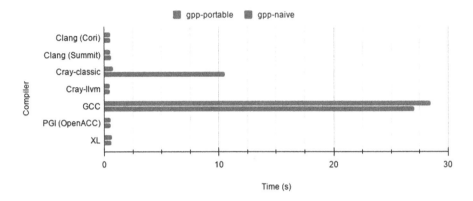

Fig. 6. Comparing execution time of gpp-portable and gpp-naive.

4.5   ToyPush

ToyPush provides an example of a larger mini-app, taken from a real-world application, that exemplifies the pattern shown in laplace. Like laplace, ToyPush launches a large number of short-running kernels. Figure 7 shows the results, using the NVTX range technique shown in Sect. 3.3. The time shown in this figure is the total time spent on each activity type within the OpenMP offloading region, as demarcated in Listing 1.1.

The relatively elevated CPU Time for XL and PGI corresponds with the total time taken to execute the mini-app, as XL and PGI were the two poorest-performing compilers by that metric. While Clang was shown in the laplace analysis to have the highest OpenMP runtime overhead, the Flang (LLVM Fortran) compiler is not able to compile this OpenMP offloading code and was thus not used in our study. Even so, Fig. 7 confirms that elevated runtime overhead impacts mini-apps derived from real-world applications, and if the future Flang compiler has a similar high overhead as in Clang, ToyPush performance would be expected to be poor.

Unlike laplace, total GPU time in ToyPush is consistent across compilers. However, note that data movement time for both Cray compilers appears elevated. nvprof profiling indicates that while the Cray compilers use pageable memory, XL and PGI used pinned memory. From discussion with compiler engineers, this is evidence that XL and PGI are performing a memory optimization, copying data to pinned memory before transfer to the GPU, broken into chunks sized to fit into pinned memory.

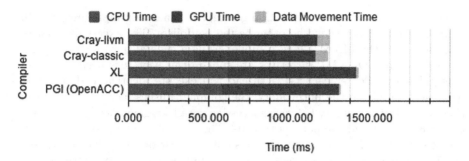

Fig. 7. Breakdown of execution time in ToyPush. Results are shown in milliseconds.

5 Discussion

In this section, we summarize and discuss the results. The results in Fig. 1 show that benchmark performance can sometimes vary by up to an order of magnitude across compilers. In Sect. 5.1 we will outline the main reasons for these performance differences and in Sect. 5.2 we will give guidelines to help application developers achieve higher performance across today's OpenMP offloading compilers.

5.1 Performance Issues Across Compilers

Three mini-apps/kernels were impacted by poor OpenMP reduction performance: laplace, gpp, babelStream-dot. We find that all the compilers generate an efficient OpenMP add reduction except for LLVM/Clang, which uses a relatively large number of general purpose compare-and-swap atomic instructions to implement an OpenMP reduction. We also find that only PGI, Cray-classic and XL compilers generate an efficient max reduction. The Cray-llvm implementation uses 4 orders of magnitude more atomic instructions than Cray-classic implying the use of a general purpose slower code path. The LLVM/Clang compiler show a similar count of compare-and-swap atomic instructions for the max and add reductions implying reuse of the compiler code. It is likely that Cray-llvm performance will improve over time as HPE incorporates more of the optimizations from Cray-classic into Cray-llvm. We hope to see OpenMP reduction performance become a priority optimization in the open-source LLVM/Clang compiler. This is because many applications, including laplace and gpp, benefit from high performance reductions. We also note that the Clang compiler, the only compiler we test on both Cori and Summit, generally shows similar performance between the two platforms.

The su3-v0 and gpp-naive mini-apps are impacted by unexpected data movement between GPU device memory and GPU registers. The su3-v0 mini-app is characterized by **teams distribute** and **parallel for** directives on separate loops. Our results show that the Clang compiler generated approximately 20x more data movement than the other tested compilers. The excess data movement is due to the LLVM/Clang compiler using a general purpose code generation path when OpenMP directives are split in this way. It is needed because it is unknown whether each GPU thread will execute identical code on independent data. Given the uncertainty, the compiler flushes memory to ensure a consistent view of memory between successive parallel regions as well as between the team master and the parallel threads. The other compilers do not generate excess data movement for su3-v0 because of compiler optimization passes, e.g. the XL compiler uses interprocedural static compiler analysis to determine that all threads in a team execute the same code [27]. The second mini-app, gpp-naive, sums data contributions over 4 nested loops. We find that Cray-classic generated code has 4 orders of magnitude more device data movement than the corresponding Clang generated code. The data movement significantly decreases in the gpp-portable version of the mini-app, where the programmer uses extra private variables to do a per-thread sum over the inner two loops before adding this sum to the OpenMP reduction variable. This indicates that it is sometimes necessary to manually exploit data reuse rather than relying on the compiler.

Finally, the Laplace mini-app is impacted by OpenMP runtime overhead for the LLVM/Clang compiler. This mini-app uses a small problem size that makes it sensitive to target region latency. The surprising observation is that target region latency can be significantly larger than kernel launch latency. In the absence of reductions, we measure $50\,\mu s$ target region time for the LLVM/Clang compiler and $7\text{--}20\,\mu s$ target region time for the proprietary compilers. This indicates that the management of an OpenMP device data environment is particularly high for the LLVM/Clang compiler. CPU profiling of this overhead to determine its cause is an area for future study.

5.2 Recommendations to Programmers

The results in the paper draw attention to some mini-apps/kernels which perform relatively well across all tested compilers. The babelStream-triad/mul/copy/add kernels and ToyPush mini-app perform consistently well. If we expand the list to include mini-apps with a median performance of >0.8 relative to the best performing compiler then it also includes babelStream-dot, gpp-portable and gpp-naive. The characteristics of these applications include

- Minimal data movement between CPU and GPU
- Combined `teams distribute parallel for` constructs
- Minimal use of OpenMP reductions
- Average GPU kernel runtime $> 50\,\mu s$

None of the mini-apps spent a large fraction of time moving data between CPU and GPU, however, this is often the biggest bottleneck in newly ported applications. We found non-negligible time spent in ToyPush and identified an interesting optimization in the PGI and XL compilers where pinned memory was used to efficiently transfer data between CPU and GPU. This could be important for users who have applications more bound by CPU-GPU data movement time than the mini-apps in our sample.

We recommend that the combined `teams distribute parallel for` constructs are used where possible. In cases where this is not possible, we draw attention to our experience with su3-v0, which has `teams distribute` and `parallel for` on separate loops. We found two different reasons for poor performance with the Cray-classic and LLVM/Clang compiler. The Cray-classic compiler performs poorly because the compiler selected a poor kernel launch configuration; we were able to improve performance by manually increasing the number of teams. The LLVM/Clang compiler performs poorly because the compiler used a general purpose code generation mode which resulted in more memory flushes.

We recommend that OpenMP reductions are used only where necessary because we found mixed performance across compilers. In addition, max reductions sometimes perform significantly worse than add reductions. One suggested method to improve reduction performance in Clang is to add the option `-fopenmp-cuda-teams-reduction-recs-num=<num>`, with `<num>` set to the number of loop iterations. However, in our experience this never led to more than a 10% speedup with any mini-app. Our experience with gpp-portable and gpp-naive is inconclusive about whether it is beneficial for the programmer to use additional private variables to reduce thread local contributions before reducing into OpenMP reduction variables. Using this technique we see significant performance improvement for the Cray-classic compiler but results in a slow down for the XL compiler. In general, we hope that compiler developers prioritize the performance of OpenMP reductions because it is a frequently used parallel pattern.

We also want to make programmers aware of OpenMP runtime overheads. We found that Laplace mini-app performance is very sensitive to OpenMP target region latency. The overhead is highest for the GCC and LLVM/Clang compiler. It should be noted that OpenMP provides a huge convenience to the programmer by enabling a single variable name to refer to data in a host and a device data environment. The cost of this convenience is that launch latencies are higher than using CUDA or some other lower-level API. Therefore programmers must make sure to send sufficient work to the GPU to justify the data transfer overhead.

We specifically recommend that, at this time, programmers use the GCC compiler primarily for correctness, not for performance. GCC was consistently a low-end outlier in our study. We hope this is temporary as OLCF is working with Mentor Graphics to improve the performance of OpenMP 5.0 features while specifically focusing on GPU offloading directives.

It is common for many developers to use roofline analysis to evaluate performance on CPU+GPU systems. We suggest that developers supplement this analysis with some additional measurements based on our experience with OpenMP mini-apps. These measurements are DRAM read/write transactions, average kernel runtime, and atomic instructions for at least 2 compilers. We have seen that DRAM read and write transactions can sometimes be much higher for some compilers than for others. If one were to rely on roofline analysis only, then it can seem like the mini-apps are achieving close to the memory bandwidth roofline, even though this would just be artifact of excessive data movement. We suggest that average kernel runtime should be measured to identify cases where runtime is less than $<50\,\mu s$. Some compilers, e.g. LLVM/Clang, would likely spend at least this much time in just CPU runtime overhead. In this case, developers should look for opportunities to fuse target region code or investigate opportunities to launch target regions asynchronously using nowait or through multiple host threads. Finally, we suggest measuring atomic instruction latency to understand if a compiler OpenMP reduction implementation is the reason for poor OpenMP reduction performance.

6   Conclusions and Future Work

This paper fills a gap in the literature, comparing OpenMP offloading compilers with multiple mini-applications derived from real-world apps and most importantly examining performance differences in detail, uncovering the specific causes for slowdowns in implementations. We do so in a way that allows us to make specific and useful recommendations to application developers. These recommendations should allow developers to avoid common performance snags found in the available compilers. Broadly, our findings regarding compiler performance show that runtime overhead in compilers tends to have a bigger negative impact on performance than architecture-specific vendor optimizations have a benefit. Without changes in the compilers, these overheads will not go away when moving to the next generation of GPUs or accelerators; the issues are more fundamental to the compilers.

Future work in this area includes studying additional GPUs, such as those produced by Intel and AMD, as well as NVIDIA Ampere (to be used in the upcoming Perlmutter system). With the addition of these hardware platforms, there is the opportunity to test additional compilers, such as Intel ICC and AMD AOMP. Further, the PGI compiler will support OpenMP offloading in the future. Beyond merely expanding coverage, a long-term goal would be creating a suite similar to the OpenMP Verification and Validation project [20], a publicly-available, well-documented suite, with a comprehensive set of kernel-only mini-apps extracted from real applications. This would provide application and compiler developers a tool for understanding the performance strengths and weaknesses of the available compilers, on various architectures, with open source and transparent tests that anyone can run.

Acknowledgements. This research used resources of the National Energy Research Scientific Computing Center (NERSC), a U.S. Department of Energy Office of Science User Facility operated under Contract No. DE-AC02-05CH11231. This research also used resources of the Oak Ridge Leadership Computing Facility, which is a DOE Office of Science User Facility supported under Contract DE-AC05-00OR22725. The authors would like to thank Doug Doerfler and Rahul Gayatri for helpful discussion about the su3 benchmark and useful research directions for this project.

References

1. Austin, B.: Nersc-10 workload analysis (data from 2018) (2020). https://portal. nersc.gov/project/m888/nersc10/workload/N10_Workload_Analysis.latest.pdf
2. Bercea, G.T., et al.: Performance analysis of OpenMP on a GPU using a CORAL proxy application. In: Proceedings of the 6th International Workshop on Performance Modeling, Benchmarking, and Simulation of High Performance Computing Systems, pp. 1–11 (2015)
3. Bertolli, C., et al.: Integrating GPU support for OpenMP offloading directives into clang. In: Proceedings of the Second Workshop on the LLVM Compiler Infrastructure in HPC, pp. 1–11 (2015)
4. Boehm, S., Pophale, S., Vergara Larrea, V.G., Hernandez, O.: Evaluating performance portability of accelerator programming models using SPEC ACCEL 1.2 benchmarks. In: Yokota, R., Weiland, M., Shalf, J., Alam, S. (eds.) ISC High Performance 2018. LNCS, vol. 11203, pp. 711–723. Springer, Cham (2018). https:// doi.org/10.1007/978-3-030-02465-9_51
5. Chandrasekaran, S., Juckeland, G.: OpenACC for Programmers: Concepts and Strategies. Addison-Wesley Professional (2017)
6. Che, S., et al.: Rodinia: a benchmark suite for heterogeneous computing. In: 2009 IEEE International Symposium on Workload Characterization (IISWC), pp. 44–54. IEEE (2009)
7. Deakin, T.: BabelStream (2020). https://github.com/UoB-HPC/BabelStream
8. Doerfler, D.: su3_bench (2020). https://gitlab.com/NERSC/nersc-proxies/su3_bench
9. Gayatri, R.: BerkeleyGW-kernels (2020). https://gitlab.com/NERSC/nersc-proxies/BerkeleyGW-Kernels-CPP
10. Gayatri, R., Yang, C., Kurth, T., Deslippe, J.: A case study for performance portability using OpenMP 4.5. In: Chandrasekaran, S., Juckeland, G., Wienke, S. (eds.) WACCPD 2018. LNCS, vol. 11381, pp. 75–95. Springer, Cham (2019). https://doi.org/10.1007/978-3-030-12274-4_4
11. Juckeland, G., et al.: SPEC ACCEL: a standard application suite for measuring hardware accelerator performance. In: Jarvis, S.A., Wright, S.A., Hammond, S.D. (eds.) PMBS 2014. LNCS, vol. 8966, pp. 46–67. Springer, Cham (2015). https:// doi.org/10.1007/978-3-319-17248-4_3
12. Juckeland, G., et al.: From describing to prescribing parallelism: translating the SPEC ACCEL OpenACC suite to OpenMP target directives. In: Taufer, M., Mohr, B., Kunkel, J.M. (eds.) ISC High Performance 2016. LNCS, vol. 9945, pp. 470–488. Springer, Cham (2016). https://doi.org/10.1007/978-3-319-46079-6_33
13. Khronos: OpenCL (2020). https://www.khronos.org/opencl/
14. Knaust, M., Mayer, F., Steinke, T.: OpenMP to FPGA offloading prototype using OpenCL SDK. In: 2019 IEEE International Parallel and Distributed Processing Symposium Workshops (IPDPSW), pp. 387–390. IEEE (2019)
15. Kokkos: kokkos/kokkos (2020). https://github.com/kokkos/kokkos
16. Koskela, T.: ToyPush (2017). https://github.com/tkoskela/toypush
17. Larrea, V.V., Joubert, W., Lopez, M.G., Hernandez, O.: Early experiences writing performance portable OpenMP 4 codes. In: Proceedings of Cray User Group Meeting, London, England (2016)
18. Martineau, M., McIntosh-Smith, S., Gaudin, W.: Evaluating OpenMP 4.0's effectiveness as a heterogeneous parallel programming model. In: 2016 IEEE International Parallel and Distributed Processing Symposium Workshops (IPDPSW), pp. 338–347. IEEE (2016)

19. Mishra, A., Li, L., Kong, M., Finkel, H., Chapman, B.: Benchmarking and evaluating unified memory for OpenMP GPU offloading. In: Proceedings of the Fourth Workshop on the LLVM Compiler Infrastructure in HPC. LLVM-HPC 2017. Association for Computing Machinery, New York (2017). https://doi.org/10.1145/3148173.3148184

20. Monsalve Diaz, J.M., Friedline, K., Pophale, S., Hernandez, O., Bernholdt, D., Chandrasekaran, S.: Analysis of OpenMP 4.5 offloading in implementations: correctness and overhead. Parallel Comput. **89**, 102546 (2019). https://doi.org/10.1016/j.parco.2019.102546

21. NVIDIA: About CUDA (2020). https://developer.nvidia.com/about-cuda

22. OLCF: Operational assessment 2019 oak ridge leadership computing facility (2020). https://www.olcf.ornl.gov/wp-content/uploads/2020/06/2019OLCF_OAR_FINAL.pdf

23. OpenACC: About OpenACC (2020). https://www.openacc.org/about

24. OpenMP: OpenMP specifications (2020). https://www.openmp.org/specifications/

25. Özen, G., Atzeni, S., Wolfe, M., Southwell, A., Klimowicz, G.: OpenMP GPU offload in Flang and LLVM. In: 2018 IEEE/ACM 5th Workshop on the LLVM Compiler Infrastructure in HPC (LLVM-HPC), pp. 1–9. IEEE (2018)

26. Sommer, L., Korinth, J., Koch, A.: OpenMP device offloading to FPGA accelerators. In: 2017 IEEE 28th International Conference on Application-specific Systems, Architectures and Processors (ASAP), pp. 201–205. IEEE (2017)

27. Tiotto, E., Mahjour, B., Tsang, W., Xue, X., Islam, T., Chen, W.: OpenMP 4.5 compiler optimization for GPU offloading. IBM J. Res. Dev. **64**(3/4), 14.1 (2019)

28. TOP500.org: June 2020 top500 (2020). https://www.top500.org/lists/top500/2020/06/

29. Vergara Larrea, V.G., Budiardja, R.D., Gayatri, R., Daley, C., Hernandez, O., Joubert, W.: Experiences in porting mini-applications to OpenACC and OpenMP on heterogeneous systems [published online ahead of print (24 April 2020)]. Concurr. Comput. Practice Exp. e5780 (2020). https://doi.org/10.1002/cpe.5780. https://onlinelibrary.wiley.com/doi/abs/10.1002/cpe.5780

30. Weinzierl, T.: ExaHyPE's OpenMP GPGPU port-lessons learned (2020). www.peano-framework.org/wp-content/uploads/2020/08/GPGPUs_Lessons_Learned.pdf

OpenACC

GPU Acceleration of the FINE/FR CFD Solver in a Heterogeneous Environment with OpenACC Directives

X. M. Shine Zhai[1], David Gutzwiller[1(✉)], Kunal Puri[2], and Charles Hirsch[2]

[1] Numeca-USA, 1044 Larkin Street, San Francisco, CA 94109, USA
{xiaomeng.zhai,david.gutzwiller}@numeca.be
[2] Numeca-International, Chaussée de la Hulpe, 189, 1170 Brussels, Belgium
{kunal.puri,charles.hirsch}@numeca.be

Abstract. OpenACC has been highly successful in adapting legacy CPU-only applications for modern heterogeneous computing environments equipped with GPUs, as demonstrated by many projects as well as our previous experience. In this work, OpenACC is leveraged to transform another Computational Fluid Dynamics (CFD) high order solver FINE/FR to be GPU-eligible. On the Summit supercomputer, impressive GPU speedup ranging from 6X to 80X has been achieved using up to 12,288 GPUs. Techniques critical to achieving good speedup include aggressive reduction of data transfers between CPUs and GPUs, and optimizations targeted at improving exposed parallelism to GPUs. We have demonstrated that OpenACC offers an efficient, portable and easily-maintainable approach to achieve fast turnaround time for high-fidelity industrial simulations.

Keywords: Heterogeneous computing · OpenACC · Computational Fluid Dynamics

1 Introduction

Heterogeneous architectures that encompass both CPUs and accelerators have become increasingly popular in the HPC community. One decade ago in 2010, only 9 supercomputers in the Top 500 list were equipped with accelerators, but the number has since grown fast, reaching 90 in 2015 and 144 in the latest June 2020 Top 500 list [1]. While various accelerators exist to accommodate different needs, such as Graphical Processing Unit (GPU), Intel Xeon Phi coprocessor and Tensor Processing Unit (TPU) etc., the best performing supercomputers tend to rely heavily on GPUs. In fact, it is the computing power from GPUs that makes exa-scale computing within reach in a manner that is economically viable and energy friendly.

To take advantage of GPUs, it is inevitable to adapt existing CPU-only applications. Since most legacy applications have been developed for a long time with rich features, it is often not practical to rewrite them in GPU-native languages,

© Springer Nature Switzerland AG 2021
S. Bhalachandra et al. (Eds.): WACCPD 2020, LNCS 12655, pp. 47–57, 2021.
https://doi.org/10.1007/978-3-030-74224-9_3

such as CUDA. On the other hand, OpenACC serves as a useful tool in porting the codes to a variety of heterogeneous systems. As a high-level directives-based programming model, OpenACC successfully helped us adapt a computational fluid dynamics (CFD) codes FINE/TURBO (which specializes in turbomachinery simulations) to be GPU-eligible, and on the Titan supercomputer at the Oak Ridge National Lab (ORNL) a 2X+ GPU speedup in time-to-solution has been demonstrated with a real-world example [4].

While a 2X+ GPU speedup was satisfactory in 2015, the growing interest from the CFD community to complete high-fidelity fluid simulations with less turnaround time has called for more aggressive GPU performance. In this paper, we present recent efforts to leverage OpenACC in achieving 6X to 80X GPU speedup on the Summit supercomputer, using up to 12,288 GPUs. In Sect. 2, we give a brief introduction of the CFD flow solver FINE/FR used for GPU adaptation. Then in Sect. 3, we discuss in detail the techniques leading to the favorable GPU performance, including reduction of data transfers between CPU and GPU, and targeted optimizations that increase the degree of exposed parallelism to GPUs. Strong scalability performance on Summit is presented in Sect. 4 before we conclude the work.

2   The FINE/FR CFD Solver

2.1   Programming Model of FINE/FR

Based on the high order flux-reconstruction (FR) method [5], FINE/FR uses compact computational stencils where the dense mathematical calculations are highly parallelizable. Such workload is well suited for GPUs as they offer significantly more hardware threads to carry out computing with high throughput. On the CPU side, FINE/FR uses a distributed-memory parallel MPI programming model, where the unstructured grids employed are statically partitioned via ParMETIS [6]. By formulation, FR method offers a high degree of accuracy in resolving fine-scale motions compared to conventional Reynolds-Averaged Navier-Stokes (RANS) solutions. As demonstrated by Fig. 1, the shock wave boundary layer interaction (SWBLI) [7], a phenomenon critical in the study of compressor stall mechanisms, is highly visible in the high order simulation in the form of lambda shocks on the upper blade surface. The bulk of the execution time is spent on the Runge-Kutta iteration loop, which contains multiple calls to BLAS matrix-matrix multiplication routines, and dozens of additional correction and calculation routines. Written in C++11 standard, FINE/FR uses object-oriented programming throughout the code, and in the core solver algorithms templatization is extensively used. Both the polymorphism and templatization pose some challenges to a neat OpenACC implementation (see discussions in Sect. 5), but good GPU speedup is not negatively affected.

2.2   Considerations for GPU Execution

Since FINE/FR is based on a legacy NUMECA framework, it is prohibitively expensive in terms of developer-hours to drastically change the underlying code

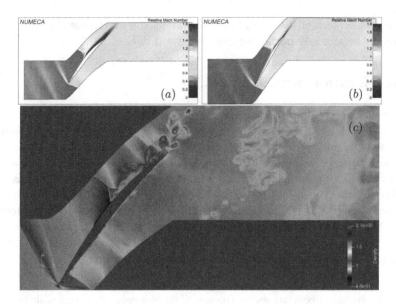

Fig. 1. Relative Mach number at (a) 50% of the blade span and (b) 95% of the blade span using the conventional RANS solution. (c) Instantaneous density snapshot using FINE/FR at polynomial order 4, where shock wave boundary layer interaction (SWBLI) in the form of the lambda-shock can be seen on the upper blade surface

structures. Yet it is useful to list a few considerations for efficient GPU executions of the FINE/FR codes.

- Linearized "flat" arrays favor data transfers: Multiple pointer indirections were natively used in FINE/FR to access array elements, and such usage is supported since OpenACC 2.0. However, current OpenACC implementation transfers each guaranteed-contiguous chunk of memory separately, which can then result in many transfers whose overheads negatively impact the performance. We have replaced multiple pointer indirections by a linearized "flat" array class which stores the data contiguously in memory. Moreover, as explained in Sect. 3.2, this array class has flags that track the last modified location of the data (CPU or GPU), thus reducing data transfers substantially.
- Sufficient workload and exposed parallelism: As GPUs become more powerful, it is important to saturate GPUs with sufficient workload to obtain significant speedup. While humongous deep learning workloads are a good candidate, we are fortunate that kernels in high-order FR methods usually have plenty of dense math to fully load GPUs too, especially at higher polynomial orders. However, at larger MPI process count, the number of cells/elements in a given partition becomes small and the amount of parallelism exposed to the GPUs is limited. Therefore, GPU speedup inevitably declines at higher MPI process count, and in fact GPU execution may no longer be cost-effective if

the exposed parallelism is too low to outweigh the overheads. Approaches to increase the exposed parallelism are discussed in Sect. 3.3.

3   Acceleration with OpenACC

Using OpenACC to adapt and accelerate FINE/FR for GPU execution is a natural choice because it does not require a rewriting of the solver in a low-level GPU language, and we had positive experience in adapting another legacy flow solver with OpenACC before [4]. Moreover, OpenACC offers good portability which allows us to conduct rapid code development on local workstations with Intel X86 architectures, and then directly ship the codes to the Summit supercomputer with IBM POWER architecture for scalability tests and production runs. Table 1 shows the system specifications used in this study. In this section, we show various optimizations with performance timed on the local workstation, and in Sect. 4, scalability performance on Summit is presented.

Table 1. System specifications for OpenACC development and large-scale testing

System	Local workstation	Summit supercomputer
CPU/Host	8 core AMD EPYC	42 core IBM POWER9 node
GPU/Device	1 Nvidia P6000	6 Nvidia V100 per node
PGI compiler	19.4	19.9
MPI library	OpenMPI 2.1.6	Spectrum MPI, 10.2.1.2

3.1   Incremental Acceleration of the Most Time-Consuming Routines

Generally, it is intuitive to identify the most time-consuming routines and look for opportunities for GPU acceleration. Figure 2 shows a representative stack trace of FINE/FR, using the low-overhead, sample-based profiler HPCToolkit [2]. As noted before, the time marching loop, composed of successive Runge-Kutta iteration steps, constitutes the bulk computation time. The most time-consuming routines were found to be BLAS calls for matrix-matrix multiplication and some thread-safe user routines, both of which are amenable to GPU acceleration via OpenACC. As a first step of acceleration, the following were implemented:

– Replace all BLAS calls with CuBLAS, the CUDA counterpart. Similar changes can be made for non-Nvidia architectures, such as AMD GPUs.
– Instrument remaining user routines (3D loops) with OpenACC pragmas for parallel execution. Minor code changes were necessary to avoid race conditions, for example, by using private variables properly.

- Offload static data such as coordinates and constants to the GPU persistently at the beginning of the program, so that they are readily available on GPUs whenever needed.
- To ensure correct results, all input and output data for each GPU-eligible routine are forced to synchronize in a conservative manner.

The first set of optimizations leads to a 1.5X GPU speedup on the local workstation, and Fig. 3 shows the updated resulting stack trace. It is clear that the time-consuming BLAS calls in the CPU run become negligible after using the CuBLAS counterpart, but numerous data transfers between CPUs and GPUs (i.e. between host and device) now dominate the execution time. In fact the amount of data transfers is excessively high due to the conservative approach of data synchronization, which always ensures that data on the host is the most up-to-date. Such an approach is useful to retain correct results when the code undergoes active development, but incurs huge waste for production runs. One alternative is to tailor-code for a particular application where unnecessary data transfers are eliminated and essential host-device communication is overlapped with computations using asynchronous queues. However the lack of generality prevents tailor-coded executables from handling various industrial settings, and would eventually demand continued investment of development efforts. As a result, a systematic and robust solution to minimize the data transfers with low maintenance cost is necessary.

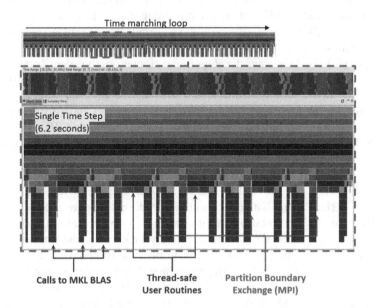

Fig. 2. Stack trace of the CPU-only execution of FINE/FR

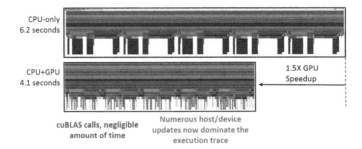

Fig. 3. Stack trace of the CPU+GPU execution of FINE/FR, after the first round of incremental optimization described in Sect. 3.1

3.2 Minimization of Data Transfer

The solution to excessive data transfers between host and device is the location-aware arrays. Essentially all major data arrays are wrapped in a container class, which, in addition to holding the linearized array data and host/device update methods, contains a "last modified" flag indicating where the array was last updated. We illustrate the usage in an example as follows.

As frame (a) of Fig. 4 shows, the conditional data synchronization between the host and device only occurs when the current access location differs from the saved last accessed location, for the particular array concerned. To ensure coherent data access, in the actual programming shown in frame (b) of Fig. 4, it is the developer's responsibility to flag the input and output arrays to GPU-eligible routines, using "sync" and "setLastAccess" calls (see "doubleValues" in the example). In this way, developers can focus on the algorithmic details of a particular routine while data transfers are automatically minimized. For example, the "doubleValues" routine is GPU-eligible but the "addValue" call is forced to run on the host. As a result, data transfers have to occur under the hood. However, should the "addValue" routine be made GPU-eligible, the approach demonstrated would completely avoid data transfers between host and device for the "addValue" call.

Since "sync" and "setLastAccess" calls are prevalent throughout the code, they are referred to as the "GPU boilerplate". The main advantage of "GPU boilerplate" is that through its consistent usage, developers are allowed to follow a "blind incremental acceleration" approach. In other words, as long as the GPU boilerplate is well in place, developers can simply tackle the most time-consuming routines one after another, and an efficient implementation with minimized data transfers would naturally follow. Moreover, new functionalities may be reliably introduced to the host-side codes, with less risk of breaking the data management in a heterogeneous workflow. Yet, it should be stressed that usage of GPU boilerplate must be mandatory for this approach to work. Moreover, existing data structures, especially Arrays of Structures (AoS), can be difficult to retrofit. We also note that other established framework exists [3] to automatically mange the issue of data locality.

```
template <typename T>
void AccArray<T>::createDevice()
{
    #pragma acc enter data copyin(this)
    #pragma acc enter data create(_data[_size])
}
template <typename T>
void AccArray<T>::deleteDevice()
{
    #pragma acc exit data delete(_data[_size])
    #pragma acc exit data delete(this)
}
template <typename T>
void AccArray<T>::updateHost()
{
    #pragma acc update host(_data[_size])
}
template <typename T>
void AccArray<T>::updateDevice()
{
    #pragma acc update device(_data[_size])
}
template <typename T>
void AccArray<T>::sync(AccessType access)
{
    if (_lastAccess != access)
    {
        if (access == HOST)
        {
            updateHost();                    Conditional
        }                                    host<->device
        else                                 updates
        {
            updateDevice();
        }
    }
}                                                      (a)
```

```
#include <iostream>
#include <AccArray.h>
using namespace std;
void doubleValues(AccArray<int>& array)
{
    array.sync(DEVICE); //GPU boilerplate
    #pragma acc parallel loop present (array)
    for (int i=0; i<array.getSize(); i++) array[i] *= 2;
    array.setLastAccess(DEVICE); //GPU boilerplate
}
void addValue(AccArray<int>& array, int adder)
{
    array.sync(HOST);
    for (int i=0; i<array.getSize(); i++) array[i] += adder;
    array.setLastAccess(HOST);
}
int main(int argc, char** argv)
{
    int size = 10;
    AccArray<int> array(size);
    array.createDevice();

    for (int i=0; i<size; i++) array[i] = i;
    array.setLastAccess(HOST);

    doubleValues(array); // GPU execution
    addValue(array,1); // CPU execution
    doubleValues(array); //GPU execution

    array.sync(HOST);
    // check results
    array.deleteDevice();                              (b)
}
```

Fig. 4. Code sample demonstrating strategies to minimize data transfer

Performance-wise, Fig. 5 confirms that minimized data transfer has significantly improved the GPU performance, leading to a 5.1X speedup compared to the CPU-only execution. In fact, the "GPU boilerplate" has reached a point where all bulk 3D data remain on the device and only 2D data along the partition boundaries needs to be transferred over MPI. Profiling shows that the time to stage the relatively small partition data is too little to warrant further optimization in data transfers, such as GPUDirect and Remote Direct Memory Access. Nevertheless, updated profiling pointed to a handful of routines with poor device performance, which motivated the continued optimization to increase the parallelism exposed to the GPUs as discussed in Sect. 3.3.

3.3   Optimization to Increase Exposed Parallelism

The high-order FR method routinely goes through a pattern of nested loops, where the outer one loops over element faces and the inner one loops over all the points per face. For unstructured grids, the number of points per face varies depending on the element type and the solution order. For example, Fig. 6 shows that the number of flux points is 12 for a quad element, while it is 9 for a triangular element. In the original CPU-only implementation as shown in frame (a) of Fig. 7, the code strictly follows the workflow, and a naive OpenACC adaptation would only parallelize the outer loop, leaving the inner loop sequential thus limiting the degree of parallelism exposed to the GPUs.

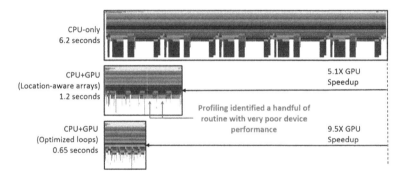

Fig. 5. Comparisons of stack trace in (from top to bottom) CPU-only run; CPU+GPU run, including incremental optimizations and minimized data transfers; CPU+GPU run, fully optimized with increased degree of exposed parallelism

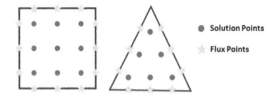

Fig. 6. Demonstration of solution points and flux points on the side of a quad and a trianglular element

Usually when the partition size is sufficiently large, each MPI process contains enough number of faces to saturate the GPUs with computations. However, limited exposed parallelism becomes an issue when the partition size is too small (at a large MPI process count), or when too few elements are treated, such as the boundary data. It turns out that collapsing the loops offers a solution to this problem. Frame (b) of Fig. 7 shows the refactored loops in a tightly nested form, where the upperbound of the inner loop is replaced by the maximum value for all iterations. The degree of exposed parallelism can increase by about one order of magnitude at the cost of some threads branching idle. Luckily there is no thread divergence issue involved, and the test on the local workstation showed a 9.5X GPU speedup compared to the CPU-only run (see Fig. 5). More detailed tuning of gang/vector parameters for the parallel loops may yield further improved performance, but the optimal parameters are likely problem-dependent. As a result, to avoid reducing portability the default parameters set by the compiler have been used.

```
#pragma acc parallel loop present(....)
for(int iFace=0; iFace<nbFaces; iFace++ )
{
    ...
    int nbPointsFace = getNbPoints(iFace);
    ...
    for (int iPoint=0; iPoint<nbPointsFace; iPoint++)
    {
        <large amounts of thread safe math>
    }
}                                        (a)
```

```
#pragma acc parallel loop collapse(2) present(....)
for(int iFace=0; iFace<nbFaces; iFace++ )
{
    for(int iPoint=0; iPoint<nbPointsFaceMax; iPoint++)
    {
        int nbPointsFace = getNbPoints(iFace);
        if (iPoint < nbPointsFace)
        {
            <large amounts of thread safe math>
        }
    }
}                                        (b)
```

Fig. 7. Code refactoring with loop collapse for improved exposed parallelism

4 Scalability of FINE/FR on Summit

The Summit supercomputer represents the state of the art heterogeneous computing architecture in the HPC community, and we are granted access through the INCITE project. Each Summit node contains 42 usable IBM POWER9 CPU cores and 6 Nvidia V100 GPUs, and applications can spawn MPI processes occupying all the 42 CPU cores per node. However as noted before, a large MPI process count gives a small partition size, which may limit the degree of parallelism exposed to the GPUs at scale. Moreover, mapping more than one CPUs per GPU, as managed by CUDA MPS server, may potentially lead to traffic congestion when all data transfers occur at the same time. As a result, Table 2 shows the strong scalability performance of the optimized FINE/FR solver on Summit where one GPU is only paired with one CPU (i.e. NbCPU = NbGPU) to maximize the partition size at large node count. Effectively it means that while GPU runs utilize all 6 GPUs per node in the GPU runs, the CPU runs only use 6 CPUs out of all the available cores as a compromise. A high resolution 8M cells mesh is used in the test with an order 3 polynomial flux reconstruction, and the effective degree of freedom is $8 \times 10^6 \times 4^3 \approx 5 \times 10^8$.

Table 2. Strong scalability using a 8×10^6 cells mesh at order 3 (5×10^8 DoF)

NbNodes	NbCPU & NbGPU	Time (s) CPU	Time (s) CPU+GPU	GPU speedup	NbCell/Partition	NbDoF/Partition
8	48	226.00	2.75	82.18	166667	10666667
16	96	117.56	1.54	76.34	83333	5333333
32	192	59.78	0.86	69.51	41667	2666667
64	384	30.59	0.58	52.74	20833	1333333
128	768	15.94	0.47	33.91	10417	666667
256	1536	7.76	0.29	26.76	5208	333333
512	3072	3.57	0.17	20.99	2604	166667
1024	6144	2.15	0.18	11.94	1302	83333
2048	12288	1.00	0.16	6.25	651	41667

The near linear scalability for the CPU-only run through the sweep reflects a well-parallelized and streamlined CPU implementation of FINE/FR and the underlying framework. The GPU implementation attains impressive speedup ranging from 6X+ to 80X+, which not only confirms that performance tuned on the local workstation is easily portable to another system, but also demonstrates the quality of the GPU optimizations. However the GPU runs see a gradual deviation from linear scalability accompanied with a reduction of GPU speedup. As shown by the "number of cells per partition" (NbCell/Partition) column in Table 2, the substantial decrease of partition size is responsible for the performance loss as the computational intensity becomes too low to saturate the GPUs. Runs with higher orders on larger meshes can increase the amount of math available to the GPUs, and relax the issue. Continued development work of further code refactoring to expose more parallelism to the device is undertaken.

The 80X+ GPU speedup needs to be interpreted with some caution. For CPU runs, it appears the Intel MKL library is highly optimized for BLAS calculations than the math library available on Summit. While for GPU runs, the CuBLAS routines give optimal performance on the Nvidia V100 cards. Therefore, the GPU speedup on Summit may not act as a perfectly fair performance comparison, and it would be interesting to revisit the GPU speedup on another supercomputer equipped with a well-tuned Intel MKL library.

5   Discussions and Conclusions

In this work, we have demonstrated the successful adaptation of FINE/FR, a Flux-Reconstruction based CFD high order solver for heterogeneous CPU/GPU architectures using OpenACC. A highlight of the present work is the use of location-aware arrays, which tracks the location where the array is last accessed. We showed that by consistently adding the "GPU boilerplate", the developers could worry less about the data synchronization between CPUs and GPUs, and focus more on introducing new features and "blindly" optimizing existing bottlenecks one by one. We also showed that increasing the exposed parallelism to GPUs added a 2X boost in parallel performance. It is encouraging to note the 9.5X GPU speedup obtained from incremental optimizations on the local workstation seamlessly translates to impressive speedup on the state of the art supercomputer, which has different CPU architectures and GPU cards, thus demonstrating the nice performance portability of OpenACC. At scale FINE/FR computations using 48 to 12,288 GPUs show favorable speedup in the range of 80X to 6X, and we stress that sufficient computation capable of saturating the GPUs is key to achieving superior GPU performance. It is worth noting that only one version of codes and executable is maintained, and overall the CPU-only execution sees neglible performance impact by the optimization.

OpenACC and the supporting PGI compiler remain actively evolving technologies, and occasionally we have to work around features that are currently not supported. For example, virtual functions and vectors in C++ had to be replaced by non-virtual ones and C-style arrays, and somewhat duplicated codes annotated by OpenACC had to exist for templated classes. Ease of use will certainly improve as OpenACC becomes more mature.

Acknowledgments. This research used resources of the Oak Ridge Leadership Computing Facility at the Oak Ridge National Laboratory, which is supported by the Office of Science of the U.S. Department of Energy under Contract No. DEAC05-00OR22725. The authors are grateful for the comments from the reviewers which have refined the presentation.

References

1. Top 500 list supercomputer statistics in June of 2010, 2015 and 2020. https://www.top500.org/statistics/list/. Accessed 20 Aug 2020
2. Adhianto, L., et al.: HPCTOOLKIT: tools for performance analysis of optimized parallel programs. Concurr. Comput. Pract. Exp. **22**(6), 685–701 (2010)
3. Ghane, M., Chandrasekaran, S., Cheung, M.S.: Gecko: hierarchical distributed view of heterogeneous shared memory architectures. In: Proceedings of the 10th International Workshop on Programming Models and Applications for Multicores and Manycores, pp. 21–30 (2019)
4. Gutzwiller, D., Srinivasan, R., Demeulenaere, A.: Acceleration of the FINE/Turbo CFD solver in a heterogeneous environment with OpenACC directives. In: Proceedings of the Second Workshop on Accelerator Programming Using Directives, pp. 1–8 (2015)
5. Huynh, H.T.: A flux reconstruction approach to high-order schemes including discontinuous Galerkin methods. In: 18th AIAA Computational Fluid Dynamics Conference, p. 4079 (2007)
6. Karypis, G., Schloegel, K., Kumar, V.: ParMETIS: parallel graph partitioning and sparse matrix ordering library (1997)
7. Touber, E., Sandham, N.D.: Large-eddy simulation of low-frequency unsteadiness in a turbulent shock-induced separation bubble. Theor. Comput. Fluid Dyn. **23**(2), 79–107 (2009). https://doi.org/10.1007/s00162-009-0103-z

Domain-Specific Solvers

Performance and Portability of a Linear Solver Across Emerging Architectures

Aaron C. Walden[1](✉), Mohammad Zubair[2], and Eric J. Nielsen[1]

[1] NASA Langley Research Center, Hampton, VA, USA
aaron.walden@nasa.gov
[2] Old Dominion University, Norfolk, VA, USA

Abstract. A linear solver algorithm used by a large-scale unstructured-grid computational fluid dynamics application is examined for a broad range of familiar and emerging architectures. Efficient implementation of a linear solver is challenging on recent CPUs offering vector architectures. Vector loads and stores are essential to effectively utilize available memory bandwidth on CPUs, and maintaining performance across different CPUs can be difficult in the face of varying vector lengths offered by each. A similar challenge occurs on GPU architectures, where it is essential to have coalesced memory accesses to utilize memory bandwidth effectively. In this work, we demonstrate that restructuring a computation, and possibly data layout, with regard to architecture is essential to achieve optimal performance by establishing a performance benchmark for each target architecture in a low level language such as vector intrinsics or CUDA. In doing so, we demonstrate how a linear solver kernel can be mapped to Intel® Xeon™ and Xeon Phi™, Marvell® ThunderX2®, NEC® SX-Aurora™ TSUBASA Vector Engine, and NVIDIA® and AMD® GPUs. We further demonstrate that the required code restructuring can be achieved in higher level programming environments such as OpenACC, OCCA, and Intel® OneAPI™/SYCL, and that each generally results in optimal performance on the target architecture. Relative performance metrics for all implementations are shown, and subjective ratings for ease of implementation and optimization are suggested.

Keywords: Programming models · Performance portability · Emerging architecture · CFD · HPC · CUDA · OpenACC · OCCA · AVX-512 intrinsics · Neon intrinsics · Arm · GPU · V100 · A100 · MI50 · Xeon Phi · SX-Aurora · ThunderX2

1 Introduction

A diverse array of new hardware architectures continues to emerge across the High Performance Computing (HPC) landscape. The application developer is faced with the considerable challenge of providing near-optimal performance across these systems. This goal requires a detailed understanding of each target

© Springer Nature Switzerland AG 2021
S. Bhalachandra et al. (Eds.): WACCPD 2020, LNCS 12655, pp. 61–79, 2021.
https://doi.org/10.1007/978-3-030-74224-9_4

architecture and some means to accommodate specific data layouts and algorithm implementations that map appropriately. Ideally, this would be achieved in a unified code base that is easily maintained. In this vein, a number of general portability approaches have recently been introduced that attempt to insulate the application developer from intricate details of the underlying hardware, yet still provide near-optimal performance on each. Unfortunately, some applications can require significant restructuring to achieve optimal performance on a particular system, which can be challenging to automate using general abstractions and run-time environments. In such cases, the developer may be required to address the needs of the underlying architecture at the application level.

The work reported here describes an ongoing effort to explore performance portability issues for the FUN3D computational fluid dynamics solver maintained at the NASA Langley Research Center [7]. FUN3D solves the Navier-Stokes (NS) equations, a system of highly nonlinear, tightly-coupled time-dependent partial differential equations. FUN3D is routinely used for a broad range of aerodynamics applications across the speed range, on both conventional x86-based systems [20], as well as GPU-based systems such as Summit at Oak Ridge National Laboratory (ORNL) [14]. FUN3D uses an implicit time-integration strategy with a node-based, finite-volume spatial discretization on general mixed-element unstructured grids. An approximate nearest-neighbor linearization of the discrete residual equations within each control volume gives rise to a large tightly-coupled system of block-sparse linear equations that must be solved at each time step. The block size is determined by the number of governing equations and may range from five to several dozen. To facilitate a practical investigation of the broad array of potential performance portability issues, the scope of the current effort is limited to optimization of the linear solver kernel used within FUN3D. The study is carried out across several familiar and emerging HPC architectures using a wide range of available programming models. While this study focuses on motifs related to linear algebra, parallel efforts aimed at unstructured-grid traversals with complex gather-scatter operations supporting flux and Jacobian construction are also ongoing but are beyond the current scope.

The block-sparse linear solver used here is memory-bound with a low arithmetic intensity. In such cases, it is critical to understand the increasingly complex memory hierarchies of today's advanced architectures and how memory bandwidth and potential reuse of computations can be effectively leveraged. For example, in the case of an NVIDIA® GPU, it is important to understand how to accommodate the application data layout and to restructure the solver algorithm to utilize the registers, shared memory, L1 and L2 caches, and DRAM effectively.

The dominant computation in the linear solver used here is a block-sparse matrix-vector product; for a broad range of applications encountered in practice, 5×5 blocks are common. The off-diagonal matrix coefficients are stored in a compressed sparse row (CSR) format [25], where two integer arrays capture the sparsity pattern of the nonzero blocks in the matrix. The nonzero blocks in a row are stored contiguously in memory, and the scalar entries within a block are stored in column-major order.

Efficient processing of such a matrix is challenging on recent CPUs offering vector architectures. Vector loads and stores are essential to effectively utilize available memory bandwidth on CPUs, and maintaining performance across different CPUs can be difficult in the face of varying vector lengths offered by each. For a sparse matrix with relatively large block sizes, it is reasonably straightforward to leverage vector loads and stores. For smaller block sizes, the computation calls for a restructuring based on the available vector length. For example, if the processor supports a vector length of 32 floating-point numbers, it is desirable to map a full dense block to a vector and organize the computation to work with this mapping. Alternatively, a CPU offering a vector length of 4 floating-point numbers may demand the mapping of a partial column of a dense block to a vector. For vector engines where the vector length may be large, say 256 elements, the data layout itself may require a substantial restructuring, leading to performance portability issues arising from different data layout requirements across architectures.

A similar challenge occurs on GPU architectures, where it is essential to have coalesced memory accesses to utilize memory bandwidth effectively. Modern GPUs support the Single Instruction Multiple Thread (SIMT) model, with a group of threads referred to as a warp (or wavefront). The dimension of this thread group can vary from one GPU to another, and the group must process consecutive memory locations to achieve coalesced memory accesses. This requires mapping the warp (or wavefront) to one or more blocks of a sparse matrix and restructuring the computation accordingly. In summary, restructuring the computation is essential, and in some cases, modifications to the underlying data layout may even be required.

The goal of this project is to assess the performance and portability of a wide variety of programming frameworks when applied to a production-scale CFD simulation code. The current work advances that goal in two ways. First, it attempts to establish, for both familiar and nascent HPC architectures, an optimal performance benchmark. In doing so, we demonstrate how a linear solver kernel can be mapped to Intel® Xeon™ and Xeon Phi™, Marvell® ThunderX2®, NEC® SX-Aurora™ TSUBASA Vector Engine, and NVIDIA® and AMD® GPUs. Second, this effort explores the ability of different programming frameworks to achieve the performance established by the benchmark for a subset of the target architectures.

2   Algorithm

For a spatial mesh containing n grid vertices, the implicit approach used within FUN3D requires frequent solutions of a large $n \times n$ linear system of equations of the form $\mathbf{A \Delta Q} = \mathbf{R}$, where $\mathbf{R}$ represents the vector of discrete residual equations, $\mathbf{A}$ is an $n \times n$ block-sparse matrix composed of dense $n_b \times n_b$ blocks, and $\mathbf{\Delta Q}$ is the vector of unknowns required to advance the nonlinear solution $\mathbf{Q}^k$

Algorithm 1 MULTICOLOR LINEAR SOLVER

1: $\Delta \mathbf{Q} = 0$
2: **for** $i \leftarrow 1$ **to** n_{iter} **do**
3:     **for** $c \leftarrow 1$ **to** n_c **do**
4:         $\Delta \mathbf{r} \leftarrow \mathbf{R_c} - \mathbf{O_c} \Delta \mathbf{Q}$
5:         $\Delta \mathbf{Q_c} \leftarrow \mathbf{D_c}^{-1} \Delta \mathbf{r}$
6:     **end for**
7: **end for**

---

at time-level k to $k+1$. The coefficient matrix $\mathbf{A}$ is based on a strictly nearest-neighbor stencil. To provide flexibility in the implementation, $\mathbf{A}$ is segregated into diagonal and off-diagonal components stored separately, namely

$$\mathbf{A} \equiv \mathbf{D} + \mathbf{O} \tag{1}$$

where $\mathbf{D}$ and $\mathbf{O}$ represent the diagonal and off-diagonal blocks of $\mathbf{A}$, respectively. The implementation in FUN3D uses 32-bit precision for $\mathbf{O}$ and $\Delta \mathbf{Q}$, while 64-bit precision is used for $\mathbf{D}$ and $\mathbf{R}$.

The block-sparse $n \times n$ matrix $\mathbf{O}$ contains nnz nonzero $n_b \times n_b$ blocks that are stored using a compressed sparse row (CSR) format [25]. Each of the n rows and columns containing $n_b \times n_b$ blocks are referred to as a *brow* and a *bcol*, respectively. Two integer arrays ia and ja are used to efficiently capture the sparsity pattern of the matrix. The array ia is a rank-1 array of size $n+1$ whose i-th entry indicates the leading nonzero block index in the i-th *brow* of $\mathbf{O}$. The array includes a fictitious $n+1$ entry to facilitate traversal of the elements through the n-th *brow*. The ja array is a rank-1 array of size nnz that provides the *bcol* index for each nonzero block. A third array is used to store the block entries proceeding from $ia(1)$ to $ia(n+1) - 1$, where the scalar entries within each $n_b \times n_b$ block are stored in column-major order.

Several linear-solver options are provided within FUN3D; the scheme most commonly used in practice is the multicolor point-implicit relaxation shown in Algorithm 1 [27,28]. In this approach, the grid vertices are grouped into n_c color groups, such that no two adjacent vertices are assigned the same color. Typical values of n_c for meshes encountered in practice are 10–15. Since the matrix $\mathbf{A}$ involves only a nearest-neighbor stencil, unknowns within a color may be updated in parallel in a Jacobi-like fashion. Color groups are processed sequentially, where solution updates within each color depend solely on the latest values of $\Delta \mathbf{Q}$ in neighboring color groups. The overall process may be repeated using n_{iter} sweeps over the entire system; a value of 15 is often observed to result in suitable convergence of the nonlinear solution.

To improve cache performance, the system of equations is renumbered such that unknowns within a color appear in consecutive order. In Algorithm 1, $\mathbf{O_c}$ and $\mathbf{D_c}$ represent submatrices of $\mathbf{O}$ and $\mathbf{D}$, respectively, for the unknowns contained in color c. $\mathbf{R_c}$ represents the nonlinear residual subvector defined by unknowns belonging to color c. Line 4 of Algorithm 1 represents a standard block-sparse matrix-vector product. Line 5 requires an inversion of each $n_b \times n_b$ block of the matrix $\mathbf{D_c}$. Here, a lower-upper (LU) decomposition of these blocks is computed beforehand and stored in place. The solution for the current block row is then obtained through a forward-backward substitution procedure. Throughout this work, the terms *block row* and *row* are used interchangeably, both referring to a matrix row of 5×5 dense blocks.

In addition to the shared-memory programming models to be presented here, the solver also accommodates an MPI message-passing approach using a standard domain-decomposition strategy for architectures with multiple sockets and/or multiple NUMA domains, as well as general multi-node, distributed-memory environments necessary for large-scale simulations. To recover the serial algorithm when using this approach, halo exchanges of partition boundary data are required at the completion of each color group before processing of the next color may proceed. To hide communication latencies associated with these halo exchanges each color group is further subdivided into values along partition boundaries and those remaining values lying entirely interior to the partition. When processing the unknowns within a color group, values along partition boundaries are determined first, then nonblocking MPI calls are used to initiate halo exchanges with neighboring partitions. Values interior to the partition are then evaluated while halo values are in flight. At the completion of the current color, each process waits for communication to complete prior to initiating the next color.

3   Architectures

Table 1 summarizes the relevant characteristics of the target architectures detailed in this section. Only characteristics relevant to the current study, which focuses on memory performance, are shown.

Table 1. Relevant characteristics of target architectures. *NUMA Domains Used* is the number of domains used (if configurable) to obtain optimal performance in this study. *Cores* refers to physical CPU cores, streaming multiprocessors, or compute units. *SP* refers to the single-precision (32-bit) vector length. *Peak Bandwidth* refers to the theoretical, as opposed to measured, peak.

	SKL	KNL	TX2	VE	V100	A100	MI50
NUMA Domains Used	2	1	2	2	1	1	1
Cores	40	64	56	8	80	128	60
Vector/Warp Length, SP	16	16	4	512	32	32	64
Peak Bandwidth, GB/s	256	485	318	1220	900	1600	1024

SKL. Intel® Xeon™ Gold 6148 (SKL) is a dual-socket CPU with 20 physical cores per socket and 2 threads per core. Its theoretical peak aggregate memory bandwidth is 256 GB/s. It has two vector units per core with 512-bit SIMD registers that support most AVX-512 instructions.

KNL. Intel® Xeon Phi™ Knights Landing (KNL) is a family of manycore x86 processors equipped with up to 72 low-frequency cores each with four hardware threads, two 512-bit vector units per core, and up to 16 GB of configurable high-bandwidth (at least 485 GB/s) 3D-stacked MCDRAM. The KNL 7230 used in this study has 64 cores. All results are run in *flat* mode, where MCDRAM is exposed as a NUMA domain, as our test case requires less than 16 GB of memory.

TX2. The Marvell® ThunderX2® (TX2) used in this study is a dual-socket processor with 28 cores per socket. The theoretical peak memory bandwidth for a dual-socket system is 318 GB/s. STREAM Triad results [26] suggest that the maximum bandwidth achievable on the system is approximately 240 GB/s, or roughly 120 GB/s for each NUMA node. The TX2 in the current study was of unknown SKU, and STREAM Triad results were at best 201 GB/s. The system can be configured to use up to four-way SMT; however, the system was configured for two-way SMT for the testing considered here.

VE. The NEC® SX-Aurora™ TSUBASA Vector Engine (VE) is a floating point coprocessor that interfaces with an x86 host through PCIe. Legacy CPU code can be compiled by the NEC® compiler and run through a seamless offloading process that does not require explicit data transfer between the host and coprocessor. Thus, legacy applications are ported and run with minimal effort. The VE is a long vector architecture with a 256×8-byte vector length, an order of magnitude beyond even the most recent AVX-512-equipped CPUs. Each VE has eight out-of-order 1.6 GHz cores and up to 48 GB of second generation High Bandwidth Memory (HBM2) with a theoretical peak aggregate memory bandwidth of 1.22 TB/s. The VE has a NUMA mode [19] that partitions its cores into two sets of four which share equal amounts of the last level cache and memory, decreasing cache conflicts. All results in the current study use this mode, which improves performance by a small amount ($\sim1\%$).

V100 and A100. NVIDIA® Tesla™ V100 and A100 are the previous and current (as of this writing) generation of NVIDIA® Tesla™ GPUs. They are equipped with 16–32 and 40 GB of HBM2 memory with approximately 900 and 1600 GB/s of theoretical peak memory bandwidth, respectively. NVIDIA® GPU hardware leverages a SIMT approach distributed across a number of streaming multiprocessors (SMs), which in turn consist of multiple cores. Threads are organized in blocks, or cooperative thread arrays, where one or more blocks run on an SM. The threads in a block are further partitioned into subgroups of 32 threads known as warps. A warp runs on eight or sixteen cores of an SM in multiple clock cycles. The NVIDIA® GPUs used in the current study are of the SXM2 variant.

MI50. AMD® Radeon™ Instinct™ GPUs, which will comprise the ORNL exascale Frontier system [24], are based on the Vega architecture and there are several models currently available, including the MI50 used in this study, MI25, and MI60. The MI50 has 60 compute units with 64 stream processors per compute unit for a total of 3,840 stream processors [4]. It has 16 GB of HBM2 memory with a theoretical peak memory bandwidth of 1,024 GB/s. From the application developer's perspective, major differences between the NVIDIA® Tesla™ V100 and the AMD® MI50 include (a) memory bandwidth (900 GB/s and 1 TB/s, respectively); (b) the warp size of 32 threads on V100 and wavefront size of 64 threads on MI50; and (c) the lack of hardware support for floating-point atomic operations on MI50.

4 Test Case

The test case used here is based on transonic turbulent flow over the semispan wing-body configuration described in Ref. [16]. The freestream Mach number is 0.85, the angle of attack is zero degrees, and the Reynolds number based on the mean aerodynamic chord is 5 million. The computational mesh consists of 1,123,718 grid vertices, 1,172,171 prisms, 3,039,656 tetrahedra, and 7,337 pyramids. This problem size is representative of the workload that would typically be placed on a single compute node in practice. For the purposes of the current study, a single linear system is extracted from an arbitrary time step during the nonlinear convergence. The linear system contains a total of 18,998,518 nonzero off-diagonal blocks, or an average of approximately 17 off-diagonal blocks per mesh vertex. Timings reported below are for 15 sweeps over the entire system.

5 Fortran Implementation

The legacy FUN3D solver implementation is written in Fortran 90 and supports both MPI [3] and MPI+OpenMP [2] programming models. In the latter case, a separate MPI rank is typically placed on each NUMA domain. The memory layout is the CSR layout described in Sect. 1 and this implementation is referred to as "Fortran (CSR)" throughout, where it is used as a performance baseline (if applicable). Figure 1 shows the loop executed for each color. The outer loop is over matrix block rows in the color. The inner loop is over blocks in a matrix row. The matrix-vector product is manually unrolled over the inner $n_b \times n_b$ dimensions and computed using scalar variables. Forward-backward substitution is also manually unrolled. This structure has been determined to perform best on common CPUs such as Intel® Xeon™ processors. When using OpenMP, parallelization occurs over block rows of the matrix. Unless stated otherwise, benchmark results use the MPI+OpenMP model with one rank per NUMA domain and one thread per hardware thread.

```
 1   !$omp parallel
 2   !$omp do private(f1,f2,f3,f4,f5,n,j,icol,istart,iend)
 3     rhs_solve : do n = start, end
 4
 5       f1 = -res( ,n)
 6       f2 = -res( ,n)
 7       f3 = -res( ,n)
 8       f4 = -res( ,n)
 9       f5 = -res( ,n)
10
11       istart = iam(n)
12       iend   = iam(n+ )-
13
14       do j = istart,iend
15         icol = jam(j)
16
17         f1 = f1   - a_off( , ,j)*dq( ,icol)
18         f2 = f2   - a_off( , ,j)*dq( ,icol)
19         f3 = f3   - a_off( , ,j)*dq( ,icol)
20         f4 = f4   - a_off( , ,j)*dq( ,icol)
21         f5 = f5   - a_off( , ,j)*dq( ,icol)
22         f1 = f1   - a_off( , ,j)*dq( ,icol)
23         f2 = f2   - a_off( , ,j)*dq( ,icol)
24         f3 = f3   - a_off( , ,j)*dq( ,icol)
25         f4 = f4   - a_off( , ,j)*dq( ,icol)
26         f5 = f5   - a_off( , ,j)*dq( ,icol)
27         f1 = f1   - a_off( , ,j)*dq( ,icol)
28         f2 = f2   - a_off( , ,j)*dq( ,icol)
29         f3 = f3   - a_off( , ,j)*dq( ,icol)
30         f4 = f4   - a_off( , ,j)*dq( ,icol)
31         f5 = f5   - a_off( , ,j)*dq( ,icol)
32         f1 = f1   - a_off( , ,j)*dq( ,icol)
33         f2 = f2   - a_off( , ,j)*dq( ,icol)
34         f3 = f3   - a_off( , ,j)*dq( ,icol)
35         f4 = f4   - a_off( , ,j)*dq( ,icol)
36         f5 = f5   - a_off( , ,j)*dq( ,icol)
37         f1 = f1   - a_off( , ,j)*dq( ,icol)
38         f2 = f2   - a_off( , ,j)*dq( ,icol)
39         f3 = f3   - a_off( , ,j)*dq( ,icol)
40         f4 = f4   - a_off( , ,j)*dq( ,icol)
41         f5 = f5   - a_off( , ,j)*dq( ,icol)
42
43       end do
44
44         f2 = f2 - a_diag_lu( , ,n)*f1
45         f3 = f3 - a_diag_lu( , ,n)*f1
46         f4 = f4 - a_diag_lu( , ,n)*f1
47         f5 = f5 - a_diag_lu( , ,n)*f1
48
49         f3 = f3 - a_diag_lu( , ,n)*f2
50         f4 = f4 - a_diag_lu( , ,n)*f2
51         f5 = f5 - a_diag_lu( , ,n)*f2
52
53         f4 = f4 - a_diag_lu( , ,n)*f3
54         f5 = f5 - a_diag_lu( , ,n)*f3
55
56         f5 = f5 - a_diag_lu( , ,n)*f4
57
58         dq( ,n) = f5 * a_diag_lu( , ,n)
59         f1 = f1 - a_diag_lu( , ,n)*dq( ,n)
60         f2 = f2 - a_diag_lu( , ,n)*dq( ,n)
61         f3 = f3 - a_diag_lu( , ,n)*dq( ,n)
62         f4 = f4 - a_diag_lu( , ,n)*dq( ,n)
63
64         dq( ,n) = f4 * a_diag_lu( , ,n)
65         f1 = f1 - a_diag_lu( , ,n)*dq( ,n)
66         f2 = f2 - a_diag_lu( , ,n)*dq( ,n)
67         f3 = f3 - a_diag_lu( , ,n)*dq( ,n)
68
69         dq( ,n) = f3 * a_diag_lu( , ,n)
70         f1 = f1 - a_diag_lu( , ,n)*dq( ,n)
71         f2 = f2 - a_diag_lu( , ,n)*dq( ,n)
72
73         dq( ,n) = f2 * a_diag_lu( , ,n)
74         f1 = f1 - a_diag_lu( , ,n)*dq( ,n)
75
76         dq( ,n) = f1 * a_diag_lu( , ,n)
77
78       end do rhs_solve
79   !$omp end do
80   !$omp end parallel
```

(a) Setup and matrix-vector product.

(b) Forward-backward substitution.

Fig. 1. Baseline FUN3D Fortran point-implicit multicolor solver.

6 Optimized Performance Benchmarks

Each section herein describes the optimization of the solver for the section's respective architecture. The resulting optimized performance is shown in Table 2.

Table 2. Optimized solver results. The time given is for 15 sweeps through the linear system in milliseconds. *% Peak Bandwidth* is the application requested bandwidth divided by the theoretical peak bandwidth for the architecture (see Table 1). Application requested bandwidth is computed by dividing the amount of bytes that must pass at least once through main memory (DRAM/MCDRAM/HBM2) by the execution time. It does not consider cache effects.

	SKL	KNL	TX2	VE	V100	A100	MI50
Optimized Time, ms	166.0	140.0	167.0	102.0	48.8	30.9	64.9
% Peak Bandwidth	78.3	52.8	62.7	26.7	75.8	67.3	51.3

6.1 Intel® Xeon™ and Xeon Phi™ Knights Landing

The Fortran solver implementation (see Sect. 5) did not perform as expected for a bandwidth-bound code given KNL's main memory bandwidth of approximately 485 GB/s. For this reason, an AVX-512 vector intrinsic [10] solver was developed. AVX-512 vector intrinsics are an abstraction just above the assembly level that can be used in a higher level language such as C++ and give the programmer fine-grained control over a thread's vector registers. There are also intrinsic instructions for memory prefetching, which is of interest in part due to the high latency of MCDRAM.

AVX-512 Intrinsic Solver. The AVX-512 intrinsic solver processes a single matrix block row, computing the matrix-vector product of each 5×5 block and the vector ΔQ, performing forward-backward substitution using the resultant vector, and storing the updated ΔQ.

The matrix-vector product is performed on chunks of three 5×5 blocks. The vector length of 512 bits holds up to 16 32-bit values. Three columns of O are loaded into a vector register with the final lane being zero. Avoiding splitting the columns across registers minimizes code complexity and load instructions while retaining over 90% vector efficiency. Corresponding values of the vector ΔQ are broadcast to 5 vector lanes in groups of three to fill a vector register using the _mm512_mask_extload_ps intrinsic. These two registers are multiplied and subtracted from an accumulator register using the _mm512_fnmadd_ps intrinsic. This process is repeated over the entire row. This produces 15 partial sums in the accumulator register. This register is permuted and summed to produce **b** in the first 5 lanes of the accumulator register. See Fig. 2a for an illustration of the matrix-vector product on a chunk of three 5×5 blocks. A remainder loop handles rows with lengths not divisible by three.

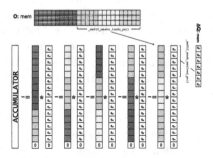

(a) Mapping from memory to vector registers and partial summation.

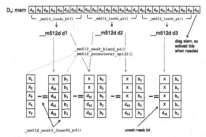

(b) Forward-backward substitution.

Fig. 2. AVX-512 solver.

Forward-backward substitution cannot achieve efficient vectorization without processing multiple matrix rows. The implementation instead attempts to minimize register usage and maximize vectorization through register permutation intrinsics. **D** is loaded once into three vector registers and permuted into operand registers as needed. Appropriate values of **b** are broadcast into multiple lanes using register permutations and summed using _mm512_mask3_fnmadd_pd with an appropriate mask. The resulting $\Delta\mathbf{Q}$ is stored to main memory. Streaming stores are not used as $\Delta\mathbf{Q}$ may reside in cache. See Fig. 2b for an illustration of AVX-512 forward-backward substitution.

The SSE and KNCI intrinsic sets contain a memory prefetch intrinsic, _mm_prefetch, with a hint argument that specifies L1, L2, and nontemporal prefetches with additional exclusivity options (for memory to be modified). The AVX-512 intrinsic solver uses this intrinsic to prefetch data for the current matrix row into L1 followed by prefetching of the next row's data into L2.

Processing three matrix rows simultaneously seems a natural extension of this algorithm that would triple vectorization efficiency of the forward-backward substitution and $\Delta\mathbf{Q}$ writes, but improved performance has not been observed for this variant.

Though originally developed for KNL, the AVX-512 intrinsic solver is also used on Intel® Xeon™ processors that support common AVX-512 instructions.

6.2 Marvell® ThunderX2®

The ThunderX2® architecture offers Neon vector units capable of supporting 128-bit vector lengths. Effective use of these vector units is challenging for block-sparse matrix-vector operations when the block size is not a multiple of the vector length. This becomes particularly difficult for a Fortran or C compiler to address in an automated fashion, and experiments confirmed that compiler-generated code yields suboptimal performance on the ThunderX2®. For this reason, an implementation based on Neon intrinsics is described here.

The ThunderX2® can be configured to use up to four-way SMT; however, the system was configured for two-way SMT for the testing considered here. Optimal performance was observed while executing a single thread per core, where the thread has access to nearly all of the resources on the core. To address NUMA issues, a hybrid approach based on the use of MPI and OpenMP is used, with one MPI rank assigned to each of the two NUMA domains.

Vectorization Using Neon Intrinsics. Processing a row of blocks for a sparse matrix-vector product involves multiplying each dense 5×5 block with a dense vector of size 5 corresponding to the column index of the block. This operation is repeated across the row, with results accumulated into a resultant vector of size 5. Since the vector length available on ThunderX2® is 128 bits, four simultaneous single-precision multiplies are possible. For $n_b = 4$, vectorization is straightforward. However, for the value of $n_b = 5$ used in the current study, each column of the 5×5 block is partitioned into two segments. The first segment consists of four elements that can be processed as a vector, while the remaining element is processed as a scalar. Figure 3a shows this partitioning and the Neon intrinsics instructions necessary to load the first four elements of each column as a vector and the remaining element as a scalar. Prefetching as shown in Fig. 3b is used to further improve performance.

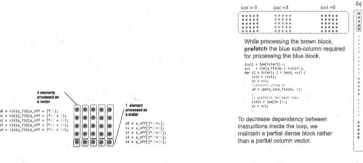

(a) Partitioning of a 5 × 5 block to enable vectorization.

(b) Prefetching **ΔQ**.

Fig. 3. ThunderX2® optimization strategies.

6.3   NEC® SX-Aurora™ TSUBASA Vector Engine

The primary challenge in achieving performance on the SX-Aurora™ is effective utilization of the long vector. The Fortran solver implementation (see Sect. 5) initially performed an order of magnitude slower on SX-Aurora™ than a conventional CPU (Intel® Xeon™ Gold 6148). To allow the NEC® Fortran compiler to vectorize over matrix block rows, the loops over rows and blocks were interchanged. Because each row may have a different number of blocks, a maximum number of blocks is computed for each color and used as the block loop range. Rows with fewer blocks than the maximum are conditionally computed and it is assumed the compiler will efficiently mask these operations when vectorizing. These changes increased the baseline performance by approximately 4.5×, but no further attempts at optimization using the original matrix memory layout were successful. In principle, one could extend the AVX-512 implementation described in Sect. 6.1 to a longer vector by vectorizing over the matrix rows. However, the AVX-512 implementation relies heavily on arbitrary register lane permutations, which are not easily done with the SX-Aurora™ instruction set.

SX-Aurora™ Optimizations Using Modified ELLPACK Memory Layout. The ELLPACK memory layout [13] regularizes a sparse matrix by treating each matrix row as having the same length, padding with zero values to extend short rows up to the maximum row length. We modified this format and applied it to the matrix **O** as follows. The dimensions of the matrix (Fortran order) become $neq \times n_b \times n_b \times l_m$ where neq is the number of matrix rows, n_b is 5 in this case, and l_m is the maximum matrix row length. For the case described in Sect. 4, l_m is 29 and the average number of rows is approximately 17, thus significant padding is introduced.

This implementation uses the interchanged loop described in the previous section. It also makes use of the NEC® Fortran compiler's **vreg** directives [18], which direct the compiler to treat local arrays as vector registers. The documentation states that packed registers (`pvreg`) of 512 floats are supported, but the `pvreg` directive did not produce working code in these experiments. An unroll directive was added to the outermost loop. The modified ELLPACK, loops, and directives improve performance by approximately another 3×, surpassing the performance of Intel® Xeon™ Gold 6148 for this kernel.

SX-Aurora™ Optimizations Using Modified SELL-C-σ Memory Layout. The SELL-C-σ memory layout [15] improves upon ELLPACK at the cost of additional complexity. Matrix rows are sorted in groups of σ and zero-padded to the maximum row length in chunks of C rows. For the case described in Sect. 4, the parameters $C = 256$ and $\sigma = n_c$ were used, where n_c is the number rows in each color group. This results in less than 2% padding. The SELL-256-n_c layout improves performance by 1.25× over the modified ELLPACK layout.

6.4 NVIDIA® Tesla™ V100 and A100 GPUs

CUDA [22] is a nonportable C++ language extension offering low-level control of NVIDIA® GPU hardware. To develop an efficient GPU implementation of the multicolor point-implicit solver, functions provided by the *cuSPARSE* [23] and *cuBLAS* [21] libraries were initially considered. The function *cusparseSbsrmv* multiplies a block-sparse matrix with a vector, and the function *cublasStrsm-Batched* solves block systems of equations by performing forward and backward substitutions using an LU-decomposition of the diagonal block. Experiments showed that this approach yields suboptimal performance for linear systems representative of those encountered in typical FUN3D simulations.

Instead, optimized CUDA implementations of these functions were developed in Ref. [28]. To perform a block sparse matrix-vector product, the proposed algorithm allocates a number of warps to process a subset of the blocks in a single row of the sparse matrix. The mapping of a warp to process a block of a sparse matrix with $n_b = 5$ is illustrated in Fig. 4. To perform forward and backward substitutions, a second kernel is invoked that assigns a single warp to process one diagonal block. Several challenges were encountered, including a variable extent of available parallelism, indirect memory addressing, low arithmetic intensity, and the need to accommodate different block sizes. To address these challenges, particular emphasis was placed on coalesced memory loads, the use of shared memory and prefetching, minimal thread divergence within warps, and strategic use of shuffle instructions available on recent hardware. Depending on the value of n_b, the new implementations realized performance gains of up to 7× over existing *cuSPARSE* and *cuBLAS* library functions [28].

6.5 AMD® Radeon™ MI50 GPU

The restructuring of the computation required for AMD and NVIDIA GPUs (see Sect. 6.4) is very similar. Since the AMD hardware calls for 64 threads per wavefront, two versions of the algorithm have been implemented: (a) one block-row per wavefront with two nonzero blocks mapped to a wavefront, and (b) two

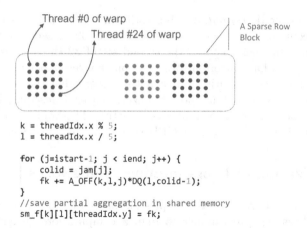

```
k = threadIdx.x % 5;
l = threadIdx.x / 5;

for (j=istart-1; j < iend; j++) {
    colid = jam[j];
    fk += A_OFF(k,l,j)*DQ(l,colid-1);
}
//save partial aggregation in shared memory
sm_f[k][l][threadIdx.y] = fk;
```

Fig. 4. Assignment of a warp to process a complete 5 × 5 block to ensure that consecutive threads of the warp load and process data from consecutive locations of device memory. The warp processes a complete row one block at a time, and aggregates partial results into a 5 × 5 block. The columns of the final aggregated block are reduced using shuffle instructions or shared memory (not shown here).

block-rows per wavefront with half of a wavefront mapped to a nonzero block of a row. We used HIP to develop an optimized implementation on AMD GPU. HIP, or Heterogeneous-Computing Interface for Portability [6], is a C++ API similar to CUDA that has been developed by AMD.

One Block-Row per Wavefront. In this algorithm, a wavefront processes two consecutive nonzero blocks of a row concurrently. Since a wavefront on the AMD GPU consists of 64 threads, 14 threads remain idle. The wavefront processes a row of the block-sparse matrix in a loop, where 2 consecutive nonzero blocks are processed by the wavefront at each iteration. The wavefront handles 50 (2 × (5 × 5)) matrix entries during each iteration. The appropriate elements of ΔQ are also loaded from the read-only data cache, multiplied by the corresponding elements of the matrix, and then results are accumulated. After completion of the loop, the 50 partial results are aggregated into an output of 5 elements. The code segment to illustrate this computation is shown in Fig. 5.

```
// nbk = 2
for (j = istart; j < iend - nbk + 1; j += nbk) {
    fk += A_OFF(k, l, j + nbki) * DQ(l, jam[j + nbki] - 1);
}
// process left over blocks
nbl = (iend - istart) % nbk;
if (tidx < nbl * nb2) {
    fk += A_OFF(k, l, j + nbki) * DQ(l, jam[j + nbki] - 1);
}
sm_f[k][l + nbki * 5][threadIdx.y] = fk;
```

Fig. 5. Code for one block-row per wavefront on AMD GPU.

Two Block-Rows per Wavefront. In this algorithm, a wavefront is assigned to process two consecutive block-rows with the first set of 32 threads (half-wavefront) processing the first block-row and the second set of 32 threads processing the second block-row. A half-wavefront processes one nonzero block of a row concurrently. Note that in this algorithm, it is not necessary that the two consecutive block-rows have an identical number of nonzero blocks. Consequently, not all of the 50 threads of a wavefront will always be active. The implementation of this algorithm is similar to the NVIDIA GPU version discussed in Sect. 6.4.

7   Optimization of Programming Frameworks

This section attempts to address the question of whether a given programming framework allows the programmer to map a computation efficiently onto an architecture and recover the performance of an optimized implementation written in a sufficiently low level language (see Sect. 6).

7.1   OpenACC

The OpenACC programming model [1] is based on the use of compiler directives and offers the potential for portable implementations across multiple GPU architectures.

NVIDIA® Tesla™ V100 and A100 GPUs. Prior development of an optimal CUDA implementation provided valuable insight in achieving a straightforward OpenACC implementation. Here, the launch parameters for each CUDA kernel were replaced with for-loops over thread blocks and the threads within each block. The sequential code annotated with OpenACC directives is shown in Fig. 6; note the similarities with the CUDA implementation shown in Fig. 4.

7.2   SYCL

SYCL is a cross-platform programming model based on C++ with support for different architectures [12]. SYCL implements a single-source, multiple compiler-passes model that allows the integration of source code for different architectures. The Intel® Data Parallel C++ (DPC++) compiler is based on SYCL with additional extensions, and provides support for a variety of OpenCL [11] devices including CPUs, FPGAs and GPUs [9]. Codeplay recently added experimental SYCL support for NVIDIA® GPUs, which avoids the use of OpenCL through use of the LLVM compiler [8]; OpenCL implementations for NVIDIA® GPUs are generally not effective due to limited NVIDIA support for OpenCL 1.2. Instead, this approach provides a plugin to DPC++ that enables compilation of SYCL code with direct CUDA support. This approach is used to evaluate SYCL performance for the NVIDIA® Tesla V100 GPU.

NVIDIA® Tesla™ V100 GPU. A SYCL implementation of the solver kernel has been developed and compiled with the Codeplay LLVM implementation.

```
#pragma acc loop gang
for (blockID = 0; blockID < nBlocks; blockID++) {
#pragma acc loop worker
    for (tidy = 0; tidy < BLOCK_DIM_Y; tidy++) {
        n = start + blockID * BLOCK_DIM_Y + tidy;
        #pragma acc loop vector
        for (tidx = 0; tidx < 32; tidx++) {
            k = tidx % 5;
            l = tidx / 5;
            if ((n < end) && (l < 5)) {
                istart = iam[n] - 1;
                iend = iam[n + 1] - 1;
                fk = 0.0;
                #pragma acc loop seq
                for ( j = istart; j < iend; j++) {
                    icol = jam[j] - 1;
                    fk += A_OFF(k, l, j) * DQ(l, icol);
                }
                // store partial terms in shared memory
                sm_f[k][l][tidy] = fk;
            }
        }
    }
    .
    .
    .
}
```

Fig. 6. Listing of sequential code with OpenACC directives. Note the similarity of this code to the CUDA code shown in Fig. 4, illustrating an identical restructuring of the computation.

The SYCL code for the solver kernel is shown in Fig. 7. Note the similarity of the SYCL implementation to the CUDA code in Fig. 4, illustrating that SYCL exposes sufficient features to achieve a CUDA-like implementation. This flexibility is useful in expressing the restructured SYCL computation in a manner necessary to achieve good performance on NVIDIA GPUs.

7.3 HIP

The *HIPify* tool provided by AMD [5] has been used to convert the CUDA kernel implementation to HIP for execution on the NVIDIA® Tesla™ V100 GPUPUs. In this experiment, the *HIPify* tool did not alter any of the original CUDA kernel code.

7.4 OCCA

OCCA is an open source approach that enables development for a variety of devices including CPUs, GPUs, and FPGAs [17]. Back-end support is provided for targets such as CUDA, OpenMP, HIP, and OpenCL. The implementation is a simple extension to C and uses "attributes" to map code to a particular device. An implementation of the solver kernel using OCCA is shown in Fig. 8. The *@outer* attribute in the outer for-loop indicates that the computation inside the loop can be parallelized, and this loop is mapped to thread blocks when using the CUDA back-end. The *@inner(0)* and *@inner(1)* loops map to the two dimensions of the thread block. The *@shared* attribute indicates the use of shared memory.

```
cgh.parallel_for<class solver_point5>(                              Execute in parallel over the range. Specify
    sycl::nd_range<2> {sycl::range<2>(gdimx, BLOCK_DIM_Y),          here the grid and thread blocks to be
    sycl::range<2>(BLOCK_DIM_X, BLOCK_DIM_Y)},                      used by CUDA backend.
    [ = ](sycl::nd_item<2> item) {

    int const tidx = item.get_local_id(0);                         Access thread and block id inside the
    int const tidy = item.get_local_id(1);                         kernel. In CUDA terminology: threadIdx.x,
    int n = start + item.get_group(0) * BLOCK_DIM_Y + tidy - 1;    threadIdx.x, and blockIdx.x
    int const k = tidx % 5;
    int const l = tidx / 5;
    float  fk;
    double f1, f2, f3, f4, f5;
    int jam0, j;
    int istart = diam[n];
    int iend = diam[n + 1] - 1;

    if ( (n < end) && (l < 5)) {
        // Loop over Non Zeros
        fk = 0;
        for (j = istart - 1; j < iend; j++) {                      Kernel code for sparse matrix vector
            jam0 = djam[j] - 1;                                    operation, where partial aggregation terms
            fk += A_OFF(k, l, j) *  DQ(l, jam0);                   are stored in the shared memory.
        }
        SM_F(k, l, tidy) = fk;
    }
}
```

Fig. 7. SYCL implementation of the solver kernel.

Note that the code shown in Fig. 8 is quite similar to the OpenACC and CUDA implementations.

8   Results

Table 3 summarizes all results. Although the vector intrinsic results are no more than $1.16\times$ higher than Fortran (CSR) for SKL and TX2, this is due to their limited memory bandwidth as the performance bottleneck. Run on a single core of SKL, the AVX-512 solver speedup over Fortran is greater than $1.5\times$. Moreover, the AVX-512 vector intrinsic solver on SKL achieves the highest percent of theoretical peak memory bandwidth among all codes in this study.

TX2 performance should not be interpreted as representative of the architecture. The machine used in this study was an anomalous prototype with seemingly lower memory bandwidth than that reported by other TX2 users.

Optimizations for SX-Aurora™ should not be considered complete. Though considerable speedup was achieved, a lower level approach such as intrinsics has yet to be implemented.

For the additional programming frameworks considered (OpenACC, HIP on V100, SYCL, and OCCA), optimized implementations were able to match (within $\sim 3\%$) the optimized benchmark for the architecture. In this work, each code is specific to a single architecture, so, for example, there are two HIP implementations, one for V100 and one for MI50. The exception to that is A100, where both the CUDA benchmark and the OpenACC version were developed and optimized for V100 (i.e., the V100 OpenACC and CUDA codes were timed on A100 without any A100-specific optimizations).

```
for (int bx = 0; bx < Nblocks; ++bx; @outer(0)) {
    @shared float sm_f[nb][nb][NTY];;
    @shared double a_diag_lu_shared[nb][nb][NTY];
    @shared float fs[nb][NTY];
    for (int ty = 0; ty < NTY; ++ty; @inner(1)) {
        for (int tx = 0; tx < NTX; ++tx; @inner(0)) {
            int const k = tx % 5;
            int const l = tx / 5;
            int n = start + bx * NTY + ty -1 ;
            if ( (n < end) && (l < 5)) {
                int istart = iam[n]-1;
                int iend = iam[n + 1] - 1;
                double fk = 0.0;
                for ( j=istart; j < iend; j++) {
                    jam0 = jam[j];
                    fk += A_OFF(k, l, j) * DQ(l, jam0 - 1);
                }
                sm_f[k][l][ty] = fk;
            }
        }
    }
}
```

Fig. 8. OCCA implementation of the solver kernel.

Table 3. Summary of results across two portability dimensions: architecture and programming model. Numeric values indicate performance relative to Fortran (CSR) on SKL (higher is better). Subjective ratings represent ease of implementation (i.e., the code runs correctly) and optimization, respectively: E – easy, M – moderate, and H – hard. Percent values show the percent of theoretical peak bandwidth achieved. Red values indicate the highest performing implementation for a given architecture, which establishes the optimized benchmark. A "-" indicates an invalid or unimplemented combination.

	SKL	KNL	TX2	VE	V100	A100	MI50
Fortran (CSR)	1.0 E/M 69.1%	0.79 E/M 31.1%	0.97 E/M 53.9%	0.53 M/M 7.7%	-	-	-
Fortran (SELL-C-σ)	-	-	-	1.84 M/H 26.7%	-	-	-
OpenACC	-	-	-	-	3.77 E/H 74.1%	5.22 E/H 57.8%	-
CUDA	-	-	-	-	3.86 M/H 75.8%	6.08 M/H 67.3%	-
HIP	-	-	-	-	3.85 M/H 75.8%	-	2.90 M/H 51.3%
SYCL for CUDA	-	-	-	-	3.79 M/H 74.5%	-	-
Vector Intrinsics	1.13 H/H 78.3%	1.34 H/H 52.8%	1.13 H/H 62.6%	-	-	-	-
OCCA	-	-	-	-	3.76 M/H 74.0%	-	2.89 M/H 51.2%

9   Conclusions and Future Work

Optimized implementations of the linear solver kernel have been established for the target architectures. For each additional programming framework considered, a solver has been implemented for a subset of the target architectures. Performance relative to the original Fortran (CSR) implementation on SKL has been reported, as well as the percent of theoretical peak bandwidth attained. Subjective ratings of implementation and optimization difficulty have been given for each combination. For this linear solver kernel, we conclude that, for the additional programming frameworks considered (OpenACC, HIP on V100, SYCL, and OCCA), it is possible to match the performance of a lower level implementation optimized specifically for the architecture. In this work, only GPU architectures were studied using the higher-level programming frameworks. Performance of a single code across multiple architectures has not been considered and that is to be the subject of future work. A more optimized benchmark for SX-Aurora^{TM} will also be developed.

Acknowledgments. The authors would like to express their appreciation to the following people for many helpful conversations pertaining to the current work: Justin Luitjens (NVIDIA Corporation), Erich Focht and Rudolf Fischer (NEC Corporation), John Linford (Arm Limited), Tim Warburton (Department of Mathematics, Virginia Tech); Noel Chalmers (AMD Incorporated), Sameer Shende (Department of Computer and Information Science, University of Oregon), and Jeff Hammond, Varsha Madananth, and Kevin O'Leary (Intel Corporation). The authors also wish to thank the High Performance Computing Incubator at the NASA Langley Research Center and the NASA Headquarters Office of Chief Engineer Research and Analysis program for providing support for this work. The support of Dr. Mujeeb Malik, Technical Lead for the Revolutionary Computational Aerosciences subproject within the NASA Aeronautics Research Mission Directorate Transformational Tools and Technologies Project, is also acknowledged.

References

1. OpenACC. https://www.openacc.org. Accessed 24 Aug 2020
2. OpenMP. https://www.openmp.org. Accessed 24 Aug 2020
3. The MPI Forum Website. http://www.mpi-forum.org. Accessed 24 Aug 2020
4. AMD Incorporated: AMD Radeon Instinct MI50 Accelerator. https://www.amd.com/en/products/professional-graphics/instinct-mi50. Accessed 24 Aug 2020
5. AMD Incorporated: HIP Porting Guide. https://rocmdocs.amd.com/en/latest/Programming_Guides/HIP-porting-guide.html. Accessed 24 Aug 2020
6. AMD Incorporated: HIP Programming Guide. https://rocm-documentation.readthedocs.io/en/latest/Programming_Guides/HIP-GUIDE.html. Accessed 24 Aug 2020
7. Biedron, R., et al.: FUN3D Manual 13.6. NASA/TM-2019-220416 (2019)
8. Codeplay: Codeplay Contribution to DPC++ Brings SYCL Support for NVIDIA GPUs. https://www.codeplay.com/portal/news/2020/02/03/codeplay-contribution-to-dpcpp-brings-sycl-support-for-nvidia-gpus.html. Accessed 24 Aug 2020

9. Intel Corporation: Intel oneAPI DPC++ Compiler (Beta). https://software.intel.com/content/www/us/en/develop/tools/oneapi/components/dpc-compiler.html. Accessed 24 Aug 2020

10. Intel Corporation: Intrinsics Guide. https://software.intel.com/sites/landingpage/IntrinsicsGuide/. Accessed 24 Aug 2020

11. Khronos Group: OpenCL. https://www.khronos.org/opencl/. Accessed 24 Aug 2020

12. Khronos Group: SYCL. https://www.khronos.org/sycl/. Accessed 24 Aug 2020

13. Kincaid, D.R., Oppe, T.C., Young, D.M.: ITPACKV 2D User's Guide, May 1989

14. Korzun, A., et al.: Effects of Spatial Resolution on Retropropulsion Aerodynamics in an Atmospheric Environment. AIAA SciTech Forum (2020)

15. Kreutzer, M., Hager, G., Wellein, G., Fehske, H., Bishop, A.R.: A unified sparse matrix data format for efficient general sparse matrix-vector multiplication on modern processors with wide SIMD units. SIAM J. Sci. Comput. **36**(5), C401–C423 (2014). https://doi.org/10.1137/130930352

16. Laflin, K.R., et al.: Data summary from second AIAA computational fluid dynamics drag prediction workshop. J. Aircraft **42**(5), 1165–1178 (2005)

17. Medina, D.S., St-Cyr, A., Warburton, T.: OCCA: A Unified Approach to Multi-Threading Languages. arXiv preprint arXiv:1403.0968 (2014)

18. NEC Corporation: SX-Aurora TSUBASA Fortran Compiler User's Guide. https://www.hpc.nec/documents/sdk/pdfs/g2af02e-FortranUsersGuide-018.pdf. Accessed 24 Aug 2020

19. NEC Corporation: SX-Aurora TSUBASA VEOS NUMA Mode Guide for Partitioning Mode. https://www.hpc.nec/documents/guide/pdfs/VEOS_NUMA_Mode4PartitioningMode_E.pdf. Accessed 24 Aug 2020

20. Nielsen, E.J., Diskin, B.: High-performance aerodynamic computations for aerospace applications. Parallel Comput. **64**, 20–32 (2017)

21. NVIDIA Corporation: cuBLAS. https://developer.nvidia.com/cublas. Accessed 24 Aug 2020

22. NVIDIA Corporation: CUDA C Programming Guide. http://docs.nvidia.com/cuda/cuda-c-programming-guide/#axzz4Hicq83a9. Accessed 24 Aug 2020

23. NVIDIA Corporation: cuSPARSE. https://developer.nvidia.com/cusparse. Accessed 24 Aug 2020

24. Oak Ridge National Laboratory: Exascale System Expected to be World's Most Powerful Computer for Science and Innovation. https://www.olcf.ornl.gov/2019/05/07/no-scaling-back-doe-cray-amd-to-bring-exascale-to-ornl/. Accessed 24 Aug 2020

25. Saad, Y.: Iterative Methods for Sparse Linear Systems, 2nd edn. Society for Industrial and Applied Mathematics, Philadelphia (2003)

26. ANANDTECH: Assessing Cavium's ThunderX2: The Arm Server Dream Realized At Last (2018). https://www.anandtech.com/show/12694/assessing-cavium-thunderx2-arm-server-reality

27. Walden, A., Nielsen, E., Diskin, B., Zubair, M.: A mixed precision multicolor point-implicit solver for unstructured grids on GPUs. In: Proceedings of the Ninth Workshop on Irregular Applications: Architectures and Algorithms, IA3 2019, Los Alamitos, CA, USA, pp. 23–30. IEEE Press (2019)

28. Zubair, M., Nielsen, E., Luitjens, J., Hammond, D.: An optimized multicolor point-implicit solver for unstructured grid applications on graphics processing units. In: Proceedings of the Sixth Workshop on Irregular Applications: Architectures and Algorithms, IA3 2016, Piscataway, NJ, USA, pp. 18–25. IEEE Press (2016)

ADELUS: A Performance-Portable Dense LU Solver for Distributed-Memory Hardware-Accelerated Systems

Vinh Q. Dang$^{(\boxtimes)}$, Joseph D. Kotulski, and Sivasankaran Rajamanickam

Sandia National Laboratories, Albuquerque, NM 87123, USA
{vqdang,jdkotul,srajama}@sandia.gov

Abstract. Solving dense systems of linear equations is essential in applications encountered in physics, mathematics, and engineering. This paper describes our current efforts toward the development of the ADELUS package for current and next generation distributed, accelerator-based, high-performance computing platforms. The package solves dense linear systems using partial pivoting LU factorization on distributed-memory systems with CPUs/GPUs. The matrix is block-mapped onto distributed memory on CPUs/GPUs and is solved as if it was torus-wrapped for an optimal balance of computation and communication. A permutation operation is performed to restore the results so the torus-wrap distribution is transparent to the user. This package targets performance portability by leveraging the abstractions provided in the Kokkos and Kokkos Kernels libraries. Comparison of the performance gains versus the state-of-the-art SLATE and DPLASMA GESV functionalities on the Summit supercomputer are provided. Preliminary performance results from large-scale electromagnetic simulations using ADELUS are also presented. The solver achieves 7.7 Petaflops on 7600 GPUs of the Sierra supercomputer translating to 16.9% efficiency.

Keywords: Dense linear systems of equations · Distributed computing · GPU acceleration · LU factorization · Performance portability

1 Introduction

Solving a dense linear equations system is one of the most fundamental problems in numerous applications in the mathematical sciences and engineering, such as biology [1], economics [2], electrical network analysis, aircraft design, radar technology [3], etc. We can find dense linear systems of equations in many applications involving the solutions of linear partial differential equations formulated as boundary integral equations (a.k.a. boundary element method) including acoustics, electrochemistry, fluid mechanics [4], elastodynamics, fracture mechanics

© Springer Nature Switzerland AG 2021
S. Bhalachandra et al. (Eds.): WACCPD 2020, LNCS 12655, pp. 80–101, 2021.
https://doi.org/10.1007/978-3-030-74224-9_5

[5], electromagnetics (method of moments) [6]. In these applications, the boundaries of the objects of interest are discretized and the integral equations are formulated into the form of $\mathbf{A} * \mathbf{x} = \mathbf{b}$ where $\mathbf{A}$ is a dense, square matrix, $\mathbf{b}$ is (are) the corresponding right-hand-side (RHS) vector(s), and $\mathbf{x}$ is (are) the unknown solution vector(s).

In order to solve $\mathbf{A} * \mathbf{x} = \mathbf{b}$, one typically uses direct solvers with lower-upper (LU) factorization, which decomposes the matrix $\mathbf{A}$ into a lower triangular matrix $\mathbf{L}$ and an upper triangular matrix $\mathbf{U}$ such that $\mathbf{A} = \mathbf{L} * \mathbf{U}$, due to its high accuracy and robustness. However, dense LU factorization has a high computational complexity of $O(N^3)$, and a memory requirement of $O(N^2)$ which might prevent itself from simulations of extremely large problems. To reduce the heavy computational burden of direct solvers, one can use iterative solvers with their computational complexities of $O(N^2\sqrt{\kappa})$ where κ is the condition number of matrix $\mathbf{A}$ [7]. Many efforts have also been devoted to further accelerate the iterative solvers. For instance, in the area of method of moments, many fast factorization schemes have been proposed in the literature to reduce the cost of matrix-vector multiplications in iterative solutions using some suitable expansions of the underlying integral kernel with some sacrifices of accuracy. Two well-known techniques are the fast multiple method (FMM) [8] and the multilevel fast multipole algorithm (MLFMA) [9] which can reduce the computational complexity to $O(N^{1.5}\sqrt{\kappa})$ and $O(N\log(N)\sqrt{\kappa})$, respectively.

Despite its high computational complexity, a direct solver often provides more robust results in cases where many iterative solvers fail to solve accurately and/or fail to converge because the system matrices are extremely ill-conditioned. Such problems, e.g. structures supporting high-quality factor resonances or extremely large problems compared to the wavelength, are very common in real-world applications. Therefore, it is essential to have efficient implementations of dense direct solvers. The dense formulation of the problem is also memory-intensive. Most problems of interest require several hundreds or thousands of nodes/GPUs to be able to fit in memory. Therefore the dense direct solvers have to be distributed-memory parallel as well. Dense LU factorizations are also compute-intensive algorithms ($O(N^3)$ FLOPS). Hence the distributed-memory, dense LU factorization has to be able to utilize the hardware accelerators available on several of the top supercomputers extremely well. This could mitigate their aforementioned computational cost and allow them to target extremely large-scale problems while providing robust solutions to applications. Some problems of interest to us such as the boundary element method applied to electromagnetics in the frequency domain [6] result in matrices $\mathbf{A}$ that are dense, complex. Hence we need to support accelerator-focused, distributed, dense LU factorizations that can handle real and complex matrices. This is a challenging problem by itself. The challenge is made even harder by the diversity in the accelerator architectures.

The current second fastest machine on the TOP500 list is the Summit system [10] located at the Oak Ridge National Laboratory (ORNL). Each compute node of the Summit system has two POWER9 CPUs and six NVIDIA V100 GPUs. The peak double-precision floating-point performance of the CPUs and the GPUs per compute node are 1.08 TFLOPS and 46.8 TFLOPS, respectively.

The third fastest supercomputer is the Sierra system [11] located at the Lawrence Livermore National Laboratory (LLNL). Each node of Sierra has two POWER9 CPUs at 1.08 TFLOPS, and four V100 GPUs for 31.2 TFLOPS. We present performance results on both these systems (Sect. 6).

We also highlight three architectures that are of interest to us in the near future. Recently, the U.S. Department of Energy has announced plans for three exascale-class supercomputers: (1) Aurora system [12], at the Argonne National Laboratory, will be delivered in 2021 with sustained performance of 1 ExaFLOPS. Each Aurora node will contain two Intel Xeon scalable processors and six X^e architecture-based GPUs; (2) Frontier system [13] at the ORNL. It will be delivered in 2021 with 1.5 ExaFLOPS of theoretical peak performance. Each Frontier node will contain one AMD EPYC CPU and four purpose-built AMD Radeon Instinct GPUs; (3) El Capitan system [14] at the LLNL is scheduled for early 2023 with 2 ExaFLOPS of theoretical peak performance. Each El Capitan node will contain one AMD EPYC CPU and four next-generation AMD Radeon Instinct GPUs. These next generation exascale HPC architectures are continuously evolving to allow for solving larger, more computationally intensive problems. At the same time, they have introduced new challenges to algorithm designs and implementations due to significantly different architectures and programming models. Therefore, it is important to develop the dense LU solver based on algorithms and implementations that are portable to future platforms.

This paper presents ADELUS, a performance-portable dense LU solver for current and next generation distributed-memory hardware-accelerated HPC platforms. ADELUS computes the LU factorization with partial pivoting and solves real/complex dense linear systems in distributed-memory using the message passing interface (MPI). The matrix is block-mapped onto the MPI tasks (either stored on CPU memory or GPU memory). In this work, the torus-wrap mapping scheme [15], which is transparent to the users, was adopted for an optimal balance of computation and communication. MPI processes compute the factorization and solve the portion of the linear system as if the matrix was torus-wrapped. A permutation operation is performed to restore the results when the solve completes. In this work, we provide performance portability by leveraging the abstractions provided in the Kokkos programming model [16] and Kokkos Kernels library [17].

The main contributions of this paper are the following:

- A parallel, dense, performance-portable, LU factorization algorithm based on torus-wrap mapping.
- An implementation of the real/complex LU factorization algorithm for traditional and accelerator-based architectures that can achieve 1.397 PFLOPS on 900 GPUs on the Summit (the world's second fastest) supercomputer. The ADELUS software is available at https://github.com/trilinos/Trilinos.
- Comprehensive analysis of the performance, scalability, and the effect of using different memory spaces on distributed-memory.

– Integration of the dense LU solver into an electromagnetic application and a demonstration of application performance on 7600 GPUs with 7.720 PFLOPS on the Sierra (the world's third fastest) supercomputer.

2 Related Work

Dense LU factorization has been studied for several decades. In this section, we list the most popular software packages which implement LU solvers related to distributed memory and/or GPU accelerators. These algorithms and implementations are the most relevant with respect to our work. Distributed-memory LU factorization implementations are available in:

– ScaLAPACK [18]: ScaLAPACK is the standard library for high-performance dense linear algebra routines on distributed-memory computers. ScaLAPACK leverages BLAS and BLACS (Linear Algebra Communication Subprograms) for extending LAPACK routines to distributed-memory computing. The library is currently written in Fortran;
– Elemental [19]: Elemental is a C++ library for distributed-memory, dense and sparse-direct linear algebra, using C++ templates for multiple precision support. It interestingly distributes the matrix by elements, which is similar to the torus-wrap mapping scheme used in ADELUS. Since 2016, the Elemental library was forked by the LLNL team under the name Hydrogen, to make use of GPU accelerators. But the supported functionality is only limited to the basic utilities and BLAS-1,-3 operations;
– DPLASMA [20]: the DPLASMA library relies on the PaRSEC [21] runtime to schedule tasks from task dependency graphs, allowing for overlapping of communication and computation. DPLASMA, however, does not support either GPU acceleration for LU solver or C++ templates.

On the other hand, node-level hardware-accelerated implementations of the LU solvers are available in:

– CULA [22]: CULA Dense is a GPU-accelerated implementation of dense linear algebra routines providing a wide set of LAPACK and BLAS capability;
– MAGMA [23]: The MAGMA library aims to provide LAPACK functionalities for heterogeneous/hybrid architectures;
– cuSOLVER [24]: The cuSOLVER library is a high-level package based on the cuBLAS and cuSPARSE libraries. It provides useful LAPACK-like features, such as dense matrix factorization and solve routines such as LU, QR, etc.

The SLATE library [25] is the state-of-the-art library that targets multi-GPU-accelerated distributed-memory systems. SLATE provides coverage of existing ScaLAPACK functionalities, both accelerated CPU-GPU based and CPU based. SLATE uses a modern C++ framework with communication-avoiding algorithms, lookahead panels to overlap communication and computation, and task-based scheduling. To the best of our knowledge, ADELUS is the first effort addressing performance portability for LU solver via Kokkos/Kokkos

Kernels libraries on distributed-memory accelerator-based architectures. We compare ADELUS' performance against some of these implementations in Sect. 6.

3   Overview of Kokkos and Kokkos Kernels

As the systems with several different accelerators become common, the need for portable programming model and portable algorithms has become critical. Portability can be addressed using several different approaches such as a directive-based approach (using OpenMP [26], OpenACC [27]), a library-based approach (using Kokkos [16], RAJA [28]) or by writing portable domain-specific languages (DSLs) if the target domain is small. Each one of these approaches has their advantages and disadvantages. In this work, we focus on the Kokkos performance-portable library to develop the dense LU solver. The primary reason we choose the library-based portable approach is due to the ability of this option to be used immediately with CPUs and GPUs effectively, and the availability of an ecosystem where options to call BLAS or LAPACK functionality is available through the Kokkos Kernels library [17].

Kokkos is a templated C++ library that uses meta-programming so users of the library will write the code once in templated C++. At compile time, these codes are mapped to an appropriate backend depending on compile time template parameters. There are backends available for OpenMP, CUDA for NVIDIA GPUs, and experimental backends for HIP for AMD GPUs, and SYCL for Intel GPUs. We use the OpenMP and CUDA backends in this work. Kokkos uses an *execution space* to determine where the computation is mapped and a *memory space* to determine where data structures live. Both aspects are key to performance. A **Kokkos View** is a data structure to store multidimensional arrays with reference counting. We utilize the Kokkos Views for storing the matrices and vectors. The matrices and vectors use different layouts depending on whether the data structures live on the CPUs or GPUs. In Kokkos library this is called **HostSpace** and **CudaSpace**. Furthermore, we also use **CudaHost-PinnedSpace** for MPI buffers for better performance. Switching the data structures from one memory space to another is controlled completely at compile time with template parameters. The solver code remains the same for all the options.

Once the data structures are in place and an execution space is chosen, the key requirement for a dense linear solver is the availability of BLAS and LAPACK functionality. Kokkos Kernels library [17] provides portable sparse-/dense linear algebra and graph kernels. It is implemented using Kokkos for portability. Kokkos Kernels also has interfaces to vendor-optimized BLAS/LAPACK when appropriate. There are custom BLAS/LAPACK kernels implemented for performance or functionality reasons as well. We depend on the Kokkos Kernels library for BLAS and LAPACK functionality on CPUs and GPUs. Kokkos Kernels uses the dense matrices stored in layouts optimized for CPU/GPU architecture and provides the BLAS/LAPACK functionality needed by the solver.

4 Application: Method of Moments for Linear Electromagnetics

An important class of problems that can be solved with the ADELUS solver are those encountered in the solution of the boundary element method applied to electromagnetics in the frequency domain. Instead of solving Maxwell's equations in 3D space via wave equations, one solves them on the boundary between regions. This class of problems solves the integral form of Maxwell's equations by using the equivalence principle and employing divergence conforming basis functions for the currents on the *surfaces* of interest [6, 29]. In the electromagnetic's community, this is termed the method of moments. The matrix produced by this numerical technique is then solved by using ADELUS. Depending on how the boundary condition is applied, it can be categorized into two main approaches: (i) Electric Field Integral Equation (EFIE) where the boundary condition is applied on the electric field; (ii) Magnetic Field Integral Equation (MFIE) where the boundary condition is applied on the magnetic field. The EFIE can be applied to both open and closed objects whereas the MFIE applies only to closed objects. Without loss of generality, we provide a brief of summary of the method of moments for EFIE in this section.

Enforcing the boundary condition at the surface, that is, the tangential component of the total field is equal to zero yields the integral equation on the surface S: $\mathcal{L}(J) = E$, where $\mathcal{L}$ is the linear operator derived from the EFIE, J is the unknown induced surface current, and E is the corresponding right hand side related to the incident field. $\mathcal{L}$ contains kernels in the form of Green's function $G(r, r') = e^{-jk|r-r'|}/|r - r'|$, where r and r' are an observation point on S and a source point on S, respectively, and k is the wave number. Let the current on S be approximated in terms of a basis function f_n defined on the surface as

$$J \cong \sum_{n=1}^{N} I_n f_n. \tag{1}$$

Typically, a triangular disretization of the surface is employed and the well-known Rao-Wilton-Glisson (RWG) function [29] is used as basis functions in (1). Applying (1) to the EFIE $\mathcal{L}(J) = E$ and using the Galerkin method to test each side of the equation yield a dense, complex, double-precision linear system

$$\sum_{n=1}^{N} \langle f_m, \mathcal{L} f_n \rangle I_n = \langle f_m, E \rangle, \tag{2}$$

where $m = 1, 2, ..., N$. Equation (3) has the form of $\mathbf{A} * \mathbf{x} = \mathbf{b}$ which can be solved by ADELUS for $\{I_n\}_{n=1}^{N}$. Elements of $\mathbf{A}$ are given by

$$A_{mn} = \int_{f_m} \int_{f_n} \left[j\omega\mu f_m \cdot f_n - \frac{j}{\omega\epsilon} \nabla \cdot f_m \nabla' \cdot f_n \right] \frac{e^{-jk|r-r'|}}{4\pi|r - r'|}, \tag{3}$$

Note that the discretization required to solve problems of interest forces the usage of capability machines that are efficient in both message passing (MPI) and threading on advanced architectures (GPUs).

To this end, ADELUS has been successfully integrated with the method of moments code EIGER [30]. This production Fortran code has been used effectively for a large class of problems and on a variety of compute platforms – its utility has been extended by the ADELUS solver. The next generation version of EIGER, GEMMA [31], is currently being developed to use the Kokkos library to increase performance in the filling of the matrix as well.

5   Parallel LU Solver Implementation

In this section, we describe the implementation of ADELUS, including the matrix implementation using Kokkos, the torus-wrap mapping scheme, and the parallel LU solver using torus-wrap mapping (factorization, backward solve and permutation).

5.1   ADELUS Interface and Storage

ADELUS accepts a dense matrix and vectors that are block-mapped to the MPI processes. The matrix is distributed to the MPI processes such that the maximum difference in the number of rows (or columns) assigned to each MPI processes is at most one. The same rule is applied to the right hand side (RHS) vectors. ADELUS provides a distribution utility function for users to calculate the workload on each MPI process based on the number of columns (rows) of the matrix, the number of the RHS vectors and the number of processes assigned to a matrix row. The function returns the number of rows, columns and RHS vectors assigned to the process, the row and column addresses of the matrix portion in the global matrix, and the row and column indices of the matrix portion in the local block map. Figure 1a shows an example of mapping the original matrix and two RHS vectors to six MPI processes with three processes per row. This utility function is used by our applications to assemble the portions of the matrix and the RHS vectors in the 2D block format correctly on each MPI process and provide them as input to ADELUS. ADELUS is then called by MPI processes taking the portions of matrix packed with RHS vectors as their inputs.

Similar to traditional dense linear solver packages, ADELUS stores its data (matrix and RHS vector portions) in each MPI process contiguously in the column-major order. For portability, the ADELUS data container is implemented by the Kokkos View with layout as Kokkos::LayoutLeft. Kokkos::LayoutLeft essentially forces Kokkos to use column-major order. The Kokkos Views are allocated either in the host memory (**HostSpace**) or in the device memory (**CudaSpace**) depending on the desired execution backend (i.e. CPU, GPU, etc.). We use Kokkos::complex so the matrices and vectors remain portable on CPUs and GPUs. For example, one can allocate a 2D view (matrix) of complex values in the host memory by:

```
Kokkos :: View<Kokkos :: complex<double >**,
            Kokkos :: LayoutLeft ,
            Kokkos :: HostSpace>
            A("A" , my_rows , my_cols );
```

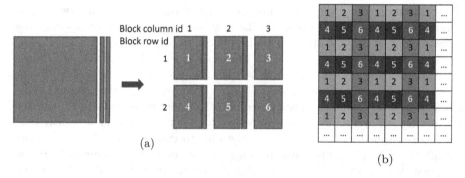

Fig. 1. ADELUS workload distribution and torus-wrap mapping for 6 MPI processes (3 processes in a row), and 2 RHS vectors. The MPI process indices are shown in the boxes: (a) Workload distribution; (b) Torus-wrap mapping.

or in the CUDA device memory by:

```
Kokkos :: View<Kokkos :: complex<double>**,
            Kokkos :: LayoutLeft ,
            Kokkos :: CudaSpace>
            A("A" , my_rows , my_cols ) ;
```

Note that the constructor takes a string that is primarily used for debugging and profiling purposes. The *my_rows* and *my_cols* are local number of rows and columns in each MPI rank. The current ADELUS solver requires the matrix and RHS vectors are packed together and computed before ADELUS is called since the forward solve is integrated with the factorization of the matrix with the RHS appended next to the matrix. This scenario is very common in the computational electromagnetics where users usually compute the matrix and the RHS vectors before calling the solvers. In order to comply with other LU solvers, we are going to provide the GETRF and GETRS functionalities separately in the upcoming ADELUS versions.

In the current version of ADELUS, the implementation is exclusive to one architecture, that is, the matrix resides in either host memory (if running on CPU backend) or device (CUDA) memory (if running on GPU backend). We plan to target a hybrid implementation where host memory and device memory are both utilized in the future versions.

5.2 Torus-Wrap Mapping

The torus-wrap mapping scheme [15] is adopted for workload distribution in ADELUS. The advantages of this mapping are each process has nearly the same workload and the process idle time is minimized. Assuming the number of MPI processes P can be factored as $P = P_r \times P_c$, where P_r is the number of processes per column and P_c is the number of processes per row, one can construct a block mapping with the block sizes of $M_p \times N_p$, where $M_p = N/P_r$ and $N_p = N/P_c$. If N

is not divisible by P_r or P_c, some processes will be assigned one more row and/or column than others. Internally, ADELUS, which uses the torus-wrap mapping scheme, assigns columns 1, $P_c + 1, 2P_c + 1$, ... to processes 1, $P_c + 1, 2P_c + 1$, ...; columns 2, $P_c + 2, 2P_c + 2$, ... to processes 2, $P_c + 2, 2P_c + 2$, ... For rows, ADELUS assigns rows 1, $P_r + 1, 2P_r + 1$, ... to processes 1, 2, ..., P_c; rows 2, $P_r+2, 2P_r+2$, ... to processes P_c+1, P_c+2, ... In other words, the column indices assigned to a MPI process constitute a linear sequence with step size P_c, and the row indices are in a sequence separated by P_r. It is not necessary to redistribute the block-mapped matrix among processes for torus-wrapped solver [15]. More specifically, a block-mapped system can be solved by a solver assuming a torus-wrapped system. In ADELUS, the solution vectors are corrected afterwards by straightforward permutations. The details are transparent to the users. Figure 1b shows an example of matrix elements torus-wrap mapped to 6 MPI processes with 3 processes per row. It should be noted that the performance of ADELUS depends on the distribution of matrix on MPI processes (i.e. the selection of P_c and P_r). It is common to choose $P_c \geqslant P_r$ for better performance. More detailed discussion is given in Sect. 6.3.

5.3   LU Solver

In ADELUS, the LU solver comprises three main steps: LU factorization+forward solve, backward solve, and permutation. We detail the algorithms for these steps in this section.

LU Factorization and Forward Solve. As the forward solve is similar to the LU factorization in terms of data use/reuse, we merge the forward solve with the factorization for performance and coding simplicity. We implement the right-looking variant of the LU factorization with partial pivoting of a dense $N \times N$ matrix. The algorithm is summarized in Algorithm 1. Each iteration in Algorithm 1 has 4 steps:

– Step 1 is to *find* the pivot. An MPI column sub-communicator is formed for the processes that own column j. Each process *finds* its own local maximum entry in the column and then exchanges within the sub-communicator for the global pivot value.
– Step 2 is to *scale* the current column j of Z with the pivot value and generate column update vector from the column j. The pivot row index and the column update vector are communicated to processes sharing the same row sets.
– Step 3 is to exchange pivot row and diagonal row. The pivot row is first *updated* and then broadcasted within each column sub-communicator. The row owner processes also send the diagonal row to processors owning the pivot row.
– Step 4 is to *update* the current column, and if saving enough columns, to *update* Z via the outer product.

Each MPI process handles its own local matrices while using Kokkos Kernels BLAS interfaces which are implemented in a simple, generic way so that the resulting code is able to run on a wide range of architectures. The BLAS interfaces enable straightforward, convenient calls to vendor library BLAS routines well-optimized for multi-threaded CPU and massively parallel GPU architectures. In this work, Kokkos Kernels calls IBM's ESSL BLAS when called with the CPU backend and calls cuBLAS when called with the CUDA backend. There are some exceptions where Kokkos Kernels calls its own implementations but they do not get used for our experiments in this work. Depending on where the data resides, Kokkos Kernels calls the right BLAS routines for the targeted backend. The BLAS operations needed in ADELUS include: (i) *KokkosBlas::iamax* for finding the local pivot entry in a column (Line 5 of Algorithm 1), (ii) *KokkosBlas::scal* for scaling the column with the inverse of pivot value (Line 10), (iii) *KokkosBlas::copy* for copying back and forth between the matrix and temporary containers (Lines 15, 20, 23, 25, 27, 31, and 33), (iv) *KokkosBlas::gemm* for updating the matrix (Lines 22, 38, and 40). It should be noted that Kokkos Kernels BLAS interfaces require Kokkos Views as their arguments. Since the algorithm needs to access subsets of the matrix, its columns and rows, we use a convenient feature of Kokkos known as the subview. A subview is a slice of a View, which is also a View of a subset of the original View. The subview types can be derived with the *auto* keyword.

Our algorithm requires only simple communication patterns consisting of point-to-point communication: MPI_Send, MPI_Recv, MPI_Irecv (Lines 16, 18, 29, 31, and 33 of Algorithm 1) and collective communication: MPI_Bcast, MPI_Allreduce (Lines 7 and 24). Furthermore, CUDA-aware MPI is exploited on GPU architectures which allows direct communication among GPUs without the need of buffering GPU data through host memory. ADELUS also has the option of using host pinned memory to buffer GPU data before communication which can be used for computer systems not having a high performance implementation of CUDA-aware MPI.

We employ the delay-updating technique (Line 39 of Algorithm 1) to take advantage of the better efficiency of level-3 BLAS *gemm* as compared to level-1 and level-2 BLAS operations. An appropriate block size parameter BLKSZ can help enhance the solver performance. A typical value of BLKSZ for CPU backend is 96 while a typical value of BLKSZ for GPU backend is 128. We determine these using several evaluations for different matrices. These numbers are used in our performance evaluation in Sect. 6. The algorithm utilizes an overlapping technique which performs column updates within a block one column at a time (Line 38). To minimize the waiting time, the algorithm attempts to do row work while waiting for a column to arrive (Line 35).

Backward Solve. In this phase, the elimination of the RHS/solution vectors is performed by the process owning the current column using the Kokkos *parallel_for* across the RHS/solution vectors (Line 4 through Line 6 of Algorithm 2). The results from the elimination step are broadcasted to all the processes within the MPI column sub-communicator (Line 7). The *KokkosBlas::gemm* is then called to update the RHS/solution (Line 8). To prepare

---

Algorithm 1. LU factorization and forward solve on MPI process p

---

Require: Matrix portion Z ($M_p \times (N_p + N_p^{rhs})$)
1 MPI process p owns row set r_p and column set c_p
   `// number of columns saved for update`
2 $colcnt = 0$
3 **for** $j = 1$ **to** N **do**
      `// Step 1: Find pivot`
4   **if** $j \in c_p$ **then**
5        $s^p \leftarrow KokkosBlas :: iamax(Z_{i \in r_p, j})$
6        $\gamma^p \leftarrow Z_{s^p, j}$
7   Exchange to compute $\gamma \leftarrow max_p \gamma^p$
8        $s \leftarrow$ row index containing the entry γ
      `// Step 2: Generate column update vector v from column j of Z`
9   **if** $j \in c_p$ **then**
10       $KokkosBlas :: scal(Z_{i \in r_p, j}, 1/\gamma)$
11   **if** $j \in r_p$ **then**
12          $Z_{j,j} = Z_{j,j} * \gamma$`// Restore diagonal`
13   **if** $s \in r_p$ **then**
14          $Z_{s,j} = Z_{s,j} * \gamma$`// Restore diagonal`
15   Copy $Z_{r_p, j}$ to $v_{r_p, colcnt}$
16   Send column $v_{r_p, colcnt}$ and s to processes sharing row set r_p
17   **else**
18   Receive s
      `// Step 3: Exchange pivot row and diagonal row, and broadcast`
      `pivot row`
19   **if** $j \in r_p$ **then**
20   Copy $[Z_{j,c_p}, v_{j,1:colcnt}]$ to $w2$
21   **if** $s \in r_p$ **then**
22       $KokkosBlas :: gemm(v_{s,1:colcnt}, u_{1:colcnt, c_p}, Z_{s, c_p})$
23   Copy $[Z_{s,c_p}, v_{s,1:colcnt}]$ to $w3$
24   Broadcast $w3$ to processes sharing column set c_p
25   Copy $w3$ to u_{s, c_p}
26   **else**
27   Receive $w3$ and copy to u_{s, c_p}
28   **if** $j \in r_p$ **then**
29   Send $w2$ to pivot owner
30   **if** $s \in r_p$ **then**
31   Receive $w2$ and copy to $[Z_{s,c_p}, v_{s,1:colcnt}]$
32   **if** $j \in r_p$ **then**
33   Copy $w3$ to $[Z_{j,c_p}, v_{j,1:colcnt}]$
34   **if** $j \notin c_p$ **then**
35   Receive $v_{r_p, colcnt}$
36   Remove j from r_p and from c_p
37    $colcnt + +$
      `// Step 4: Column update and outer product update`
38    $KokkosBlas :: gemm(v_{r_p, j}, u_{s, 1:colcnt}, Z_{r_p, 1:colcnt})$
39   **if** $colcnt = BLKSZ$ **then**
40       $KokkosBlas :: gemm(v_{r_p, 1:colcnt}, u_{1:colcnt, c_p}, Z_{r_p, c_p})$

---

for the next iteration, the newly-computed RHS/solution vectors are sent to the processes to the left processes.

Algorithm 2. Backward Solve on MPI process p

Require: Matrix portion Z $(M_p \times (N_p + N_p^{rhs}))$
1   MPI process p owns row set r_p
2   **for** $j = N$ **downto** 1 **do**
3   **if** $j \in r_p$ **then**
      `// Do an elimination step on the column and the rhs owned by`
      `   process` p
4   **for** $k = 1$ **to** N_p^{rhs} **do**
5        $u1(k) \leftarrow Z_{j,N_p+k}/Z_{j,j}$
6        $Z_{j,N_p+k} \leftarrow u1(k)$
7   Broadcast $u1$ in the column communicator
     `// Update rhs`
8     $KokkosBlas :: gemm(Z_{r_p,j}, u1(:, N_p^{rhs}), Z_{r_p, N_p^{rhs}})$
9   Send rhs to the processes on the left
10   Receive rhs from the processes on the right

Permutation. Since the torus-wrap mapping scheme is assumed by the solver while the input matrix is not torus-wrapped, a permutation of the solution vectors must be carried out to "unwrap the results". The algorithm is quite straightforward. Each process that owns local solution vectors creates a temporary buffer for global solution vectors. The permutation simply involves Kokkos *parallel_fors* to fill the local vectors to the right locations in the global vectors and an MPI_Allreduce to collectively update the change from other processes.

6 Results

6.1 Experimental Setup

We use the second and the third the fastest supercomputers in the world at the time of this writing for all our experiments, namely the Summit system at the Oak Ridge Leadership Computing Facility (OLCF), and the Sierra system at the Lawrence Livermore National Laboratory.

The Summit system contains 256 racks, each with eighteen IBM POWER9 AC922 nodes, for a total of 4,608 nodes. Each node contains two POWER9 CPUs, twenty two cores each, and six V100 GPUs. Each node has 512 GB of DDR4 memory. Each GPU has 16 GB of HBM2 memory. The processors within a node are connected by NVIDIA's NVLink 2.0 interconnect. Each link has a peak bandwidth of 25 GB/s (in each direction). The nodes are connected with a Mellanox dual-rail enhanced data rate (EDR) InfiniBand network. The software environment used for the experiments on Summit includes GNU Compiler Collection (GCC) 7.4.0, CUDA 10.1.243, Engineering Scientific Subroutine Library (ESSL) 6.2.0, Spectrum MPI 10.3.1.

The Sierra system has 240 racks, each with eighteen IBM POWER9 AC922 nodes, for a total of 4,320 nodes. Each node contains two POWER9 CPUs, twenty two cores each, and four V100 GPUs. Each node has 256GB of DDR4 memory. Each GPU has 16GB of HBM2 memory. The processors within a node are connected by NVIDIA's NVLink 2.0 interconnect. The nodes are connected with a Mellanox dual-rail enhanced data rate (EDR) InfiniBand network. The software environment used for the experiments on Sierra includes GNU Compiler Collection (GCC) 7.2.1, CUDA 10.1.243, Engineering Scientific Subroutine Library (ESSL) 6.2.0, Spectrum MPI 10.3.0.

In the next two sections, we demonstrate the performance of ADELUS. First, we investigate the performance of ADELUS solving matrices that are randomly generated on the Summit system. This is reasonable as the performance of the solver is not very different based on the values. The pivoting is the only part that could get affected. Random matrices always require pivoting, making this a good test. Second, we integrate ADELUS into a production application code, EIGER, and demonstrate performance on the linear systems from the electromagnetic application on the Sierra system.

6.2   Performance Results with Randomly-Generated Matrices

In our performance analysis, we run experiments to solve for a linear equation system with a single RHS vector and the matrix size is increased as we increase the hardware resource[1]. Note that the single right hand side problem is typically harder than multiple right hand side problem as there is one less dimension to exploit the parallelism. For the GPU backend, ADELUS runs with one MPI rank per GPU. For the CPU backend, there are three possible MPI rank configurations on the Summit system: (a) 1 MPI rank per node (42 cores each), 1 MPI rank per sockets (21 cores each), or 6 MPI ranks per node (7 cores each). It should be noted that the CPU computation time, which heavily depends on BLAS operations (in which matrix-matrix multiply for the matrix updates is the most time-consuming), dominates the total CPU execution time, as compared to the communication time. We observe that the best performance for CPU execution is reached by assigning all 42 cores for 1 MPI rank. Consequently, in our experiments, ADELUS runs with one 42-core CPU node per MPI process on CPU backend. Since the CPU memory capacity is much larger than the GPU memory capacity, it is difficult to determine a fair comparison scheme between the two backends. In this study, we opt to use the memory occupied by a matrix $(N \times N)$ represented in double complex precision in a single GPU as the baseline. As the number of MPI processes increases, the problem (i.e. matrix) sizes are increased so that each MPI process holds the same amount of matrix portion

---

[1] The driver code used for our ADELUS experiments can be found in https://github.com/trilinos/Trilinos/tree/master/packages/adelus/example.

$(N \times N)$. The baseline $N \times N$ matrix is chosen with $N = 27,882$ which takes 77.7% of 16GB GPU memory. The matrix sizes will be $N \times N$, $2N \times 2N$, ... $\sqrt{p}N \times \sqrt{p}N$, where p is the number processes, in the 1, 4 (2 processes/row), ... p processes ($\sqrt{p}$ *processes/row*), respectively. It is noted that ADELUS can handle non-square matrix portion in MPI processes. In Sect. 6.3, we will show the results of different matrix distributions. For the GPU backend, we test MPI data buffers allocated in GPU memory (CUDA-aware MPI) and host pinned memory.

Load Balancing Verification. We first look at the execution time on all MPI processes by picking the matrix size of $6N \times 6N$ running on 36 GPUs. Figure 2a and Fig. 2b show the timing breakdowns for each of the 36 processes (36 GPUs) for the factorization step in solving the $167,292 \times 167,292$ problem in double complex precision using CUDA-aware MPI and host pinned memory, respectively. The timing breakdown includes the time to find the local maximum entries (called *Local pivot*), the time for MPI communication (called *Msg passing*), the time for internal copying (called *Copying*), and the time for updating matrix (called *Update*). In case of using host pinned memory for MPI, the time for copying back and forth between the device memory and the host pinned memory is included (called *Host pinned mem copying*). It is observed that the workload (computation and communication) is almost perfectly balanced across all the MPI processes while the process idle time is kept minimized due to the torus-wrap mapping scheme. When host pinned memory is used for MPI communication, extra memory copying is explicitly made which results in the increase in the total time. We observe that the communication and the update contribute the most to the total time and the communication time is even higher than the update time (1.47x–1.6x) with this certain problem size on 36 MPI processes. This ratio is expected to increase as more nodes are added. More analysis of the communication and computation is provided in the following sections.

CPU vs. GPU Performance Comparisons. The CPU and GPU (using host pinned memory for MPI) computation time and communication time where the problem size varies from $N \times N$ on 1 MPI rank to $10N \times 10N$ matrix using 100 MPI ranks are shown in Fig. 3a and Fig. 3b, respectively. The computation time is defined by subtracting the overhead associated with MPI communication from the total execution time. We can make several observations. First, when a single GPU is compared to 42 cores of the CPU we see a speedup of 4.9 (23 s vs 113 s). Second, the GPU times increase from 23 s to 361 s from 1 rank to 100 ranks as the problem size grows one hundred times while the FLOPS grow $O(N^3)$. For the same increase in problem size, the CPU times increase from 113 to 1368 from 1 rank to 100 ranks. Finally, we can see that the GPU total execution for the $10N \times 10N$ problem on 100 processes outperforms the CPU total execution with a speedup factor of 3.8. The ratios between communication and computation are 0.43 (CPU) and 2 (GPU) for the $10N \times 10N$ problem. As processing larger problems (by more MPI processes), communication overhead increases. This

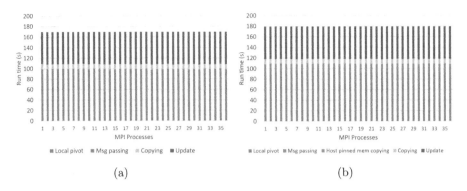

Fig. 2. Timing breakdowns of the factorization for the $6N \times 6N$ problem on 36 MPI processes using: (a) Cuda-aware MPI; (b) host pinned memory for MPI.

communication overhead is mostly contributed by the cost of broadcasting pivot rows (Line 24 of Algorithm 1–Factorization and Forward solve) and the cost of exchanging rhs vectors to left and right processes (Line 9 and 10 of Algorithm 2–Backward solve). It is noted that messages sizes depend on the size of the matrix portion held by MPI processes and these two communications happen at each iteration of the algorithms. In spite of that, CPU computation is still the dominant component in the total CPU time. However, in GPU computation, due to the fact that the computation cost is reduced by the increased parallelism on the GPUs, the communication overhead now becomes the bottleneck.

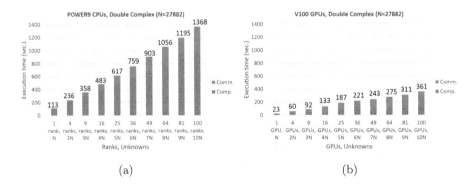

Fig. 3. ADELUS execution times (double complex precision): (a) CPU execution times. The total CPU time at $10N \times 10N$ is1368 s; (b) GPU execution times with host pinned memory. The total GPU time at $10N \times 10N$ is 361s.

Performance Comparison with DPLASMA and SLATE. ADELUS is compared against the two state-of-the-art solver packages DPLASMA [20] (CPU runs) and SLATE [25] (CPU and GPU runs) on the Summit system using the

GESV testing programs accompanied with the packages. It should be highlighted that IBM XL C/C++ Compiler 16.1.1 is used to build DPLASMA, instead of GCC 7.4.0. For building SLATE, we use GCC 6.4.0 and ESSL 6.1.0, Netlib SCALAPACK 2.0.2. DPLASMA's and SLATE's testing programs have multiple tuning parameters. We identify the values of these parameters that could give the best performance on CPUs and GPUs. We do not use the default parameters for these third party libraries. We tune them to obtain the best performance out of them. We also compare against DPLASMA despite it having the option to do only incremental pivoting while ADELUS does partial pivoting. More specifically, for DPLASMA with *GESV* functionality on CPUs, a square tile with size of 352 is exploited. For SLATE on CPUs, we can achieve the best performance with $nb = 320$, $ib = 32$, *panel_threads* $= 4$. For SLATE's *GESV* runs on GPUs, the best performance can be obtained with $nb = 640$, $ib = 32$, *panel_threads* $= 1$. Figure 4a gives GFLOPS performance of the three packages solving up to a $10N \times 10N$ matrix with 100 MPI processes on CPUs. The CPU performance of ADELUS is higher than the CPU performance of SLATE (43 TFLOPS vs. 38 TFLOPS). This can be explained by the fact that SLATE uses OpenMP threads explicitly for multitasking on individual tiles and uses BLAS functions in sequential mode while ADELUS uses multi-threaded BLAS routines. DPLASMA, with its use of the PaRSEC runtime to overlap computation and communication and to dynamically manage and schedule tasks, outperforms ADELUS on CPUs (57 TFLOPS vs. 43 TFLOPS). However, it is noted that DPLASMA does not provide the *GESV* testing with partial pivoting. We use the incremental pivoting for DPLASMA runs instead.

The GPU performance comparison is given in Fig. 4b. Due to the job time limit on Summit, we could not run SLATE further than 144 GPUs solving for $12N \times 12N$ matrix. As we can see, ADELUS delivers superior performance compared to SLATE. Using 144 GPUs, ADELUS can be 4.57x faster than SLATE. Two possible reasons are the use of batched BLAS calls on batches of tiles in SLATE and extra complication of layout translation for row swapping operation in SLATE's GPU acceleration. Another possible reason for the inferior performance of SLATE could be the overhead of simultaneous OpenMP tasks issuing MPI communications during the panel factorization in the SLATE's LU implementation. ADELUS can achieve 1,316 TFLOPS (1.3 PFLOPS) when running on 900 GPUs. To the best of our knowledge, this is the first time that a complex, dense LU solver can reach PFLOPS performance. We also emphasize that ADELUS code is identical for the CPU and GPU evaluations except one template parameter and any use of host pinned memory for MPI communication.

Scalability Analysis. In order to investigate the scalability of ADELUS, we compare how the GFLOPS performance improves with more GPUs or more nodes while we increase the matrix size, as shown in Fig. 5a. Scalability is defined as the normalized GFLOPS performance of multiple MPI processes in reference to GFLOPS performance of a single MPI process. In general, the increase of communication overhead results in less than ideal scalability in both CPU and

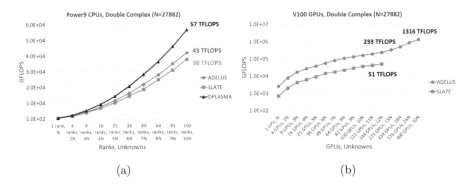

(a)   (b)

Fig. 4. GFLOPS (double complex precision): (a) ADELUS vs. DPLASMA and SLATE on Power9 CPUs; (b) ADELUS vs. SLATE on V100 GPUs.

GPU runs. It can be seen that ADELUS running on CPUs scales more closely to the theoretical ideal scalability than ADELUS running on GPUs. This can be explained by the increase in the communication costs on GPUs. This also demonstrates the fact that ADELUS clearly benefits from GPU acceleration. However, notice that the GPU's single GPU GFLOPS was already quite high, so the increase in communication cost shows in the scaling plots.

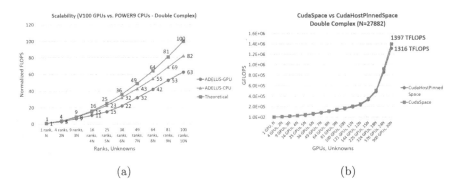

(a)   (b)

Fig. 5. ADELUS (a) Scalability (double complex precision, host pinned memory for MPI is used with GPU backend); (b) CUDA backend execution - using Cuda Host Pinned Memory vs Cuda Memory for MPI.

MPI Buffers on Different Memory Spaces. ADELUS has an option which allows one to choose whether using host pinned memory as MPI buffers or use CUDA-aware MPI during the communication. Figure 5b shows the GFLOPS performance of the GPU execution with respect to the increase of problem size. Both memory spaces, namely CudaSpace and CudaHostPinnedSpace, can attain

Table 1. ADELUS Solver Performance on Large Scale EM Simulations. Nodes are shown. Number of GPUs are four times the number of nodes.

Order(N)	Nodes	Solve Time(s)	TFLOPS	Procs/Row (P_c)
226,647	25	240.5	1291	10
1,065,761	310	1905.1	1694.5	31
1,322,920	500	6443.9	958.1	20
1,322,920	500	2300.2	2684.1	50
1,322,920	500	2063.6	2991.9	100
2,002,566	1200	3544.1	6042.6	100
2,564,487	1900	5825.2	7720.7	80

performance above 1000 TFLOPs. Using CUDA-aware MPI can improve the performance by 6% since we do not need to explicitly buffer data on host memory before or after calling the MPI function.

6.3   Performance Results from Large-Scale EM Simulation

We demonstrate ADELUS performance on a real computational science application on 100 GPUs to 7600 GPUs by integrating it with the electromagnetics simulation code EIGER (written in Fortran). Several numerical simulations were performed on the Sierra platform available at the LLNL using EIGER coupled with the ADELUS solver. The performance results are shown in Table 1. The NVIDIA GPUs were used in the solve and since there are 4 GPUs per node, the number of MPI processes is four times the number of nodes.

A number of observations can be made from Table 1. First, the performance of the solver increases with the number of nodes. *ADELUS reaches 7.72 Petaflops when using 7600 GPUs.* This translates to 16.9% of theoretical double precision floating point performance if we only account for computation cost in theory. In addition, the performance is affected by the distribution of the matrix on the MPI processes. This is revealed by the 1.3 million unknown problem where assigning more processes per row yields higher performance. We hypothesize this is due to the reduction of communication cost of broadcasting pivot rows during partial pivoting (Line 24 of Algorithm 1). However, the overhead of communicating rhs vectors to left and right processes (Line 9 and 10 of Algorithm 2) also contributes to the total performance. As we have more processes per row, this communication overhead in the backward solve increases. Therefore, we observe the performance improvement of 1.1x when going from 50 processes/row to 100 processes/row (as compared to 2.8x going from 20 processes/row to 50 processes/row) in Table 1. The selection of the number of processes per row P_c (and the number of processes per column P_r) for best performance is heuristic-based and should be a compromise to both the aforementioned communication overheads.

It is common to choose $P_c = P_r$ or P_c slightly greater than P_r for an acceptably good performance. Not shown in Table 1 is the per process performance and for the problems and distributions used has a maximum value of 1.5 Tflops/rank.

7 Conclusions and Future Work

In this paper, we present a parallel, dense, performance-portable, LU solver based on torus-wrap mapping and LU factorization algorithm. Using the portability provided by Kokkos, the solver can be portable to CPUs and GPUs. The performance evaluation of ADELUS is demonstrated on the Summit system, in which it achieves 1.397 PFLOPS on 900 GPUs. It is shown that, the GPU execution outperforms the CPU execution (with 42 cores) in terms of speedup by a factor of 3.8. We also demonstrate the integration of the ADELUS solver into an electromagnetic application achieving a performance of 7.720 PFLOPS on 7600 GPUs when solving a problem of 2.5M unknowns on the Sierra system. ADELUS scalability on the GPU backend could be resolved by exploiting more computation-communication overlapping techniques and/or by using a mixed-precision algorithm that allows factoring a matrix in low-precision and using iterative refinement to eventually achieve a high precision results. The mixed-precision approach can accelerate data transfers rates, reduce communication overhead, especially on GPUs with Tensor Core support for mixed-precision. Another issue that remains to be resolved is the limitation of the GPU memory. Since ADELUS execution is exclusive to one memory space, when the problem size exceeds the GPU memory limit, more GPUs need to be accommodated. One possible solution to overcome this limitation is a hybrid implementation where both CPU and GPU resources are fully utilized. Our future investigation would address these issues.

Acknowledgment. Sandia National Laboratories is a multimission laboratory managed and operated by National Technology and Engineering Solutions of Sandia, LLC., a wholly owned subsidiary of Honeywell International, Inc., for the U.S. Department of Energy's National Nuclear Security Administration under contract DE-NA-0003525.

Appendix: Data Availability Statement

Summary of the Experiments Reported

We ran ADELUS tests on ORNL's Summit supercomputer using gcc 7.4.0, CUDA 10.1.243, IBM ESSL 6.2.0, IBM Spectrum MPI 10.3.1, and on LLNL's Sierra supercomputer using gcc 7.2.1, CUDA 10.1.243, IBM ESSL 6.2.0, IBM Spectrum MPI 10.3.0, as described in the paper. For comparison, we also ran DPLASMA's gesv and SLATE's gesv tests on ORNL's Summit supercomputer. For DPLASMA tests, IBM XL C/C++ Compiler 16.1.1 was used. For SLATE tests, we used gcc 6.4.0, IBM ESSL 6.1.0, Netlib SCALAPACK 2.0.2.

Artifact Availability

Software Artifact Availability: All author-created software artifacts are maintained in a public repository under an OSI-approved license.

Hardware Artifact Availability: There are no author-created hardware artifacts.

Data Artifact Availability: There are no author-created data artifacts.

Proprietary Artifacts: No author-created artifacts are proprietary.

List of URLs and/or DOIs where artifacts are available: DOI: 10.6084/m9.fig share.13647497

Baseline Experimental Setup, and Modifications Made for the Paper

Relevant hardware details: Summit, Sierra

Operating systems and versions: Red Hat Enterprise Linux Server 7.6 running Linux kernel 4.14.0

Compilers and versions: GCC 7.40, GCC 7.2.1, IBM XL C/C++ Compiler 16.1.1, GCC 6.4.0

Applications and versions: ESSL 6.2.0, ESSL 6.1.0, SCALAPACK 2.0.2

Libraries and versions: CUDA 10.1.243, Spectrum MPI 10.3.1, Spectrum MPI 10.3.0

Key algorithms: gesv

References

1. Gross, E., Harrington, H.A., Rosen, Z., Sturmfels, B.: Algebraic systems biology a case study for the WNT pathway. Bull. Math. Biol. **78**(1), 21–51 (2015). https://doi.org/10.1007/s11538-015-0125-1
2. Judd, K.L.: Numerical Methods in Economics. MIT Press, Cambridge (1998)
3. Larson, R.: Elementary Linear Algebra, 8th edn. Cengage Learning, Boston (2017)
4. Wrobel, L.C., Aliabadi M.H.: The Boundary Element Method. Applications in Thermo-Fluids and Acoustics, vol. 1. Wiley, New York (2002)
5. Wrobel, L.C., Aliabadi M.H.: The Boundary Element Method. Applications in Solids and Structures, vol. 2. Wiley, New York (2002)
6. Harrington, R.F.: Field Computation By Moment Method. Wiley-IEEE Press, New York (1993)

7. Bettencourt, M.T., Zinser, B., Jorgenson, R.E., Kotulski, J.D.: Performance portable sparse approximate inverse preconditioner for EFIE equations. In: Proceedings of the International Conference on Electromagnetics in Advanced Applications (ICEAA), Verona, pp. 1469–1472. IEEE (2017)

8. Coifman, R., Rokhlin, V., Wandzura, S.: The fast multipole method for the wave equation: a pedestrian prescription. IEEE Antennas Propag. Mag. **35**(3), 7–12 (1993)

9. Song, J.M., Chew, W.C.: Multilevel fast multipole algorithm for solving combined field integral equations of electromagnetic scattering. Microw. Opt. Technol. Lett. **10**, 14–19 (1995)

10. Oak Ridge Leadership Computing Facility. https://www.olcf.ornl.gov/summit/. Accessed 20 Apr 2020

11. Livermore Computing Center - High Performance Computing. https://hpc.llnl. gov/hardware/platforms/sierra/. Accessed 23 May 2020

12. Argonne Leadership Computing Facility. https://aurora.alcf.anl.gov/. Accessed 20 Apr 2020

13. Oak Ridge Leadership Computing Facility. https://www.olcf.ornl.gov/frontier/. Accessed 20 Apr 2020

14. Smith, R.: El Capitan supercomputer detailed: AMD CPUs & GPUs to drive 2 exaflops of compute. AnandTech, March 2020. https://www.anandtech.com/show/ 15581/el-capitan-supercomputer-detailed-amd-cpus-gpus-2-exaflops

15. Hendrickson, B.A., Womble, D.E.: The torus-wrap mapping for dense matrix calculations on massively parallel computers. SIAM J. Sci. Comput. **15**(5), 1201–1226 (1994)

16. Edwards, H.C., Trott, C.R., Sunderland, D.: Kokkos: enabling manycore performance portability through polymorphic memory access patterns. J. Parallel Distrib. Comput. **74**(12), 3202–3216 (2014)

17. Kokkos Kernels. https://github.com/kokkos/kokkos-kernels. Accessed 26 Aug 2020

18. Blackford, L.S., et al.: ScaLAPACK Users' Guide. SIAM, Philadelphia (1997)

19. Poulson, J., Marker, B., Van de Geijn, R.A., Hammond, J.R., Romero, N.A.: Elemental: a new framework for distributed memory dense matrix computations. ACM Trans. Math. Softw. (TOMS) **39**(2), 13 (2013)

20. Bosilca, G., et al.: Flexible development of dense linear algebra algorithms on massively parallel architectures with DPLASMA. In: Proceedings of IEEE International Symposium on Parallel and Distributed Processing Workshops and Phd Forum (IPDPSW), Shanghai, pp. 1432–1441. IEEE (2011). https://bitbucket.org/ icldistcomp/dplasma

21. Bosilca, G., et al.: DAGuE a generic distributed DAG engine for high performance computing. Parallel Comput. **38**(1–2), 37–51 (2012)

22. Humphrey, J.R., Price, D.K., Spagnoli, K.E., Paolini, A.L., Kelmelis, E.J.: CULA: hybrid GPU accelerated linear algebra routines. In: Proceedings of SPIE Defense and Security Symposium (DSS) (2010)

23. Dongarra, J., et al.: Accelerating numerical dense linear algebra calculations with GPUs. In: Kindratenko, Volodymyr (ed.) Numerical Computations with GPUs, pp. 3–28. Springer, Cham (2014). https://doi.org/10.1007/978-3-319-06548-9_1

24. cuSOLVER library. https://docs.nvidia.com/cuda/cusolver. Accessed 26 Aug 2020

25. Gates, M., et al.: SLATE: design of a modern distributed and accelerated linear algebra library. In: Proceedings of the International Conference for High Performance Computing, Networking, Storage and Analysis, Denver, pp. 1–18. ACM (2019). https://bitbucket.org/icl/slate

26. Dagum, L., Menon, R.: OpenMP: an industry standard API for shared-memory programming. IEEE Comput. Sci. Eng. **5**(1), 46–55 (1998)
27. Farber, R.: Parallel Programming with OpenACC. Morgan Kaufmann Publishers Inc., San Francisco (2016)
28. Beckingsale, D.A., et al.: RAJA: portable performance for large-scale scientific applications. In: Proceedings of 2019 IEEE/ACM International Workshop on Performance, Portability and Productivity in HPC (P3HPC), Denver, pp. 71–81. IEEE (2019). https://github.com/LLNL/RAJA
29. Rao, S.M., Wilton, D.R., Glisson, A.W.: Electromagnetic scattering by surface of arbitrary shape. IEEE Trans. Antennas Propagat. **30**(3), 409–418 (1982)
30. Wilton, D.R., et al.: EIGER: a new generation of computational electromagnetics tools. In: Proceedings of ElectroSoft: Software for Electrical Engineering Analysis and Design, San Miniato, Italy, pp. 28–30 (1996)
31. Langston, W.L., et al.: Massively parallel frequency domain electromagnetic simulation codes. In: Proceedings of International Applied Computational Electromagnetics Society Symposium (ACES), Denver, CO (2018)

Author Index

Printed in the United States
by Baker & Taylor Publisher Services